Polluted Promises

Polluted Promises

Environmental Racism and the Search for Justice in a Southern Town

Melissa Checker

New York University Press
New York and London

NEW YORK UNIVERSITY PRESS
New York and London
www.nyupress.org

Library of Congress Cataloging-in-Publication Data
Checker, Melissa.
Polluted promises : environmental racism and the search for justice in a southern town / Melissa Checker.
p. cm.
Includes bibliographical references and index.
ISBN-13: 978-0-8147-1657-1 (cloth : alk. paper)
ISBN-10: 0-8147-1657-1 (cloth : alk. paper)
ISBN-13: 978-0-8147-1658-8 (pbk. : alk. paper)
ISBN-10: 0-8147-1658-X (pbk. : alk. paper)
1. Environmental justice—Georgia—Augusta. 2. Hazardous waste sites—Environmental aspects—Georgia—Augusta—Case studies. 3. Racism—United States. 4. African Americans—Social conditions. 5. Social justice—United States. I. Title.
GE235.G4C46 2005
363.7'009758'64—dc22 2005006793

New York University Press books are printed on acid-free paper, and their binding materials are chosen for strength and durability.

Manufactured in the United States of America
c 10 9 8 7 6 5 4 3 2 1
p 10 9 8 7 6 5 4 3 2 1

This book is dedicated to the memory of Lillian Rosen Goldberg, to the memory of Reverend Robert Louis Oliver Jr. (who would have helped me write it), and to Reverend Charles Utley and Deacon Arthur Smith Jr., who have taught me more about faith, courage, and perseverance than I could ever put into words.

Contents

Acknowledgments

Exactly six years and two days before sitting down to write these acknowledgments, I arrived in Augusta, Georgia, to begin this project. Since then, the number of generous souls who have come forward to help has unceasingly astounded and inspired me. To all those who have reaffirmed my faith in all that is good, I can only begin to express my deepest gratitude.

Certainly, the activists and residents of Hyde Park made this project both possible and a labor of love. They not only welcomed me to the neighborhood and fed me barbeque but also allowed me to become part of their lives and patiently encouraged me throughout the project's long years. In particular, with his characteristic goodwill and courage, Charles Utley took a chance on an unknown outsider, and he continues to offer his enthusiasm, wisdom, support, and warm heart. Arthur Smith welcomed me to Hyde Park with open arms; I am fortunate to be among those who benefit from his unbridled generosity and friendship. Irene Sapp and Deborah Jackson accepted a stranger into their workplace and put up with my endless questions. Vanessa Sapp showed me the ropes in Hyde Park. Her daughters, Vanesha and Ashley, both befriended and inspired me, and her father, Ernest Sapp, made me chitterlings. Terence Dicks provided me with a more radical vision of Augusta activism, as well as his friendship. Melvin Stewart and Bobby Truitt also lent me their valuable insights and good-heartedness. Brenda Utley, Robert and Viola Striggles, David Jackson, and Johnnie Mae Brown warmly offered information, contacts, and consideration.

Toward the end of this project, several angels descended to take it under their wings, and for the hours they have spent with it I extend my deepest and eternal thanks. Laura Helper-Ferris (who could not be more aptly named) cared for the manuscript as if it were her own, helping it grow by leaps and bounds through her magical blend of intelligent commentary and kind encouragement. I cannot thank her enough for enthusiastically coming through in crunch after crunch and for being my good friend. Michael Mosier, Esq., offered his time and his legal expertise, saving a stranger's skin. Maryl Levine came at the beginning, with her wonderful photographic eye, and then again at the finish, with lustrous cover ideas. My editor at NYU Press, Ilene Kalish, ventured out on several limbs for a project she believed in, and Frank Miller stepped up to the plate at the last minute. Katherine Lambert-Pennington, Jennifer Prough, and Greg Milner offered editing expertise and condiments right up to the bitter end, and Jane and Stan Herring were on hand for last-minute bolstering.

At the project's beginning, there was Owen Lynch, to whom I am forever grateful for his intelligent comments and consistent support, mentoring, generosity, patience, and overall hand-holding. I also warmly thank Constance Sutton for her intellectual guidance and her consistently abundant moral support. Steven Gregory pointed me to the topic of environmental justice back in 1995 and offered astute advice. Karen Blu was exceedingly generous with her encouragement, time, and incisive input, as was Bonnie McCay.

This project also owes much to Maggie Fishman and Jennifer Patico, who have edited reams of pages, listened to endless whining, and offered endless patience in return. Robert Brimhall, cartographer extraordinaire, donated ample time to this cause. I am also indebted to Julie Rogers, who is largely responsible for making appendix B happen, and Rebecca Puckett, who did some of the dirty work. Connie Tucker spent many hours making sure I got it right. Steve Albert, Peter Zablieskis, Jessica Winegar, Wendy Leynse, Shalini Shankar, and anonymous reviewers also contributed greatly to the editing process.

William McCracken, Lisa Milot, Alyssa Senzel, and Josh Watson offered valuable (and greatly needed) legal help. Richard Moore, Gordon Blaker, Hameed Malik, Lillian Wan, Doris Bradshaw, Zack Lyde, Evell Ballard, Juanita Burney, Frank Rumph, and Gary Grant were also good enough to support and enhance this project. Augusta State University's Robert Johnston, Kim Davies, Ernestine Thompson, and the Moral Maximalists have been generous collaborators. Brad Owens and Julee Bode welcomed me to another part of Augusta and took the time to get to know Hyde Park. In advance, I thank Jeff Howe and Stephanie Foxman for helping to launch this project in a very public way.

A National Science Foundation Doctoral Dissertation Improvement

Award (no. 9806988) funded my field research. A New York University Dean's Dissertation Award and a Morris K. Udall Environmental Policy Foundation Dissertation Fellowship enabled me initially to write up that research.

Last, but certainly not least, I thank my parents, Ruth and Armand, and my sisters and best friends, Alison and Jill, all of whom have worried with me over every obstacle and celebrated every victory, and who share in every success. Finally, I am so grateful to James R. Garfield, who continues to pat my hair (from near and far) when I want to tear it out.

This book concerns a legal case, *Jordan, et al. v. Southern Wood, et al.* At the time of this writing, claims are still pending that preclude the case from being fully resolved by a court of law. In the summer of 2003, portions of a draft manuscript were sent to the attorney representing the defendant in this lawsuit, who pointed to some additional research that has been valuable to this manuscript. At the same time, the book's intentions were questioned, and significant concerns over its biases were expressed. As a result, the following statement will clarify this book's intent:

The purpose of this book is socioanthropological insight, not legal analysis. Accordingly, legal issues are reported to the extent that they factor into the interpretations of the people depicted and socioanthropological analysis. The people portrayed in this book may make statements of their own that are recounted as evidence of their opinions, not necessarily factual evidence in support of their legal position.

One

You Can Run, but You Can't Hyde

On the afternoon of his thirty-ninth birthday, Arthur Smith Jr., a tall, handsome African American[1] man with graying hair and an easy smile, rushed into the library of the Clara E. Jenkins Elementary School. It was 3:30 P.M., and Smith was late. His last engagement, one of four community organization meetings or activities he had attended that day, had run over. Although it was only March, in Augusta, Georgia, heat and humidity had already descended, and small beads of sweat stood out on Smith's brow. Catching his breath and assuming one of his trademark grins, Smith strode into the room. Seated around three small tables, twelve African American fourth and fifth graders turned their heads.

"Mr. Arthur Jr.'s here," exclaimed Frank, a bright fourth grader.[2]

Smith walked to the library's windows and stood before the children. Behind him stretched the school's well-trodden brownish green field, with its rusting jungle gym. Framed on two sides by long ditches, the field marked the beginning of Augusta's Hyde Park neighborhood, home to approximately two hundred African American families. If you followed the ditches, which lined both sides of the neighborhood's seven streets, you would see rows of small, mostly one-story shotgun-style homes surrounded by ample lawns. Some of these houses were freshly whitewashed cottages. Others were covered in peeling paint and leaned on their foundations. Between almost every home and its yard was a porch, usually set up with chairs and often filled with families and neighbors. Some yards bloomed with daisies, chrysanthemums, and lilacs, but since the early 1990s, none included a single patch of vegetables.

Clearing the spring's pollen from his throat, Smith asked, "What's contamination?"

"Pollution," answered Cherise.

Smith nodded and then explained,

> Pollution. That's right. Poison is another word for it. Contamination is poison. What did Dr. King do during the civil rights movement? He marched. Why was he a marcher? Because everybody's an American no matter what race, color, or creed you are. He was marching because the poor in this country are left out. To show the country how to live up to its constitution, "We the people" means everything. Forty years ago, nobody thought that racism in the South could be broken. What is racism?

Fifth grader Shanequa Jones replied, "Somebody don't like the color of your skin."

Smith nodded once again. Having spent most of his life a block from the home of Shanequa's grandmother (where Shanequa spent half the week), Smith had known her since the day she was born. He continued,

> That's right. Somebody don't like you for the color of your skin. See that's what happened in Hyde Park and Aragon Park. Somebody said that we wasn't human. But the God who serves, sits high and looks low. So, things are happening in this area and it's all for you. It's for your college education; it's to make sure that you have a job when you grow up. It's to make sure you get everything you're entitled to.

Pausing for a moment, Smith glanced out of the window. "Has anybody ever noticed the signs that say playing in ditches is hazardous to your health?" he asked.

Lorenzo Thomas, a shy fourth grader who lived a few houses away from Smith on Willow Street, answered quietly, "On Willow Street."

Smiling briefly at Lorenzo, Smith continued,

> You know, when I was a little boy we used to catch tadpoles in the ditches and save them and trade them. Yeah, we did all that. But young people, do not go in the ditches out here. Ditches here are highly polluted. Contaminants can get into your feet. Understand you got to take care of you. We need y'all healthy.

Arthur Smith's speech to the children of Jenkins Elementary School that warm spring day traversed subjects from contamination to civil

"Caution" sign by Leona Street playground, 2004. Photo by author.

rights to democracy to educational and economic opportunity to childhood memory and back to contamination.[3] The multitude of topics that Smith wrapped into one discussion of his neighborhood's history is no accident. For Smith and his fellow activists in Hyde Park, all these things are inextricably linked to each other and to a history fraught with discrimination and struggle.

Like many Americans, I spent most of my life thinking of the environment as primarily a white middle-class issue. Saving whales, recycling, and preserving forests seemed hardly to affect the lives of poor and minority people, especially since these people tend to live in urban areas. My own activist endeavors centered on housing and other urban social justice concerns. Then, while searching for a research topic on a major issue facing America's cities, I read Robert Bullard's *Dumping in Dixie* (1990), which clearly and explicitly outlined how racist institutions caused people of color to bear a disproportionate burden of our nation's toxic waste. Bullard made me realize that I had been dead wrong in not considering the environment a social justice issue. Rather, it is inextricably linked to our country's history of racism.

I began my research on environmental justice activism in Brooklyn, New York, where I worked with a group of Hasidic, Latino, and

African American activists who had come together to fight the installation of a fifty-two-foot incinerator in their neighborhood. Over the course of several months, I sat at community meetings and interviewed activists, listening as they described how many of their children, relatives, and neighbors were sick with strange forms of cancer and other diseases. I marveled at how the New York City government could slate this neighborhood, which already had two highways running through it and a Superfund site on its perimeter, for an incinerator.[4] In this case, local activists eventually won the support of enough politicians to stop the plan.[5] As I further researched environmental justice and searched for a location for a longer research project, I realized that the Brooklyn situation was not uncommon—but the residents' success in fighting it was.

I thus set out to investigate (1) how environmental organizing has both changed and been incorporated into urban African American activism, and (2) what I, as an anthropologist, could do to participate in environmental justice. Hyde Park turned out to be a perfect location for answering both questions: it has been polluted in both an ecological and social sense since its inception as a neighborhood, and its residents have been fighting that pollution equally as long. For not only are Hyde Park's ecological resources (air, water, and soil) contaminated, but its social resources (access to decent jobs, housing, schools, and police protection) are also contaminated due to a history of discrimination against African Americans. The resilience and fortitude that I found among Hyde Park's residents, despite all these experiences, truly enlightened and inspired me. It is that perseverance that I have tried to depict and explore in this book.[6]

Welcome to Hyde Park

Just after World War II, Hyde Park's first residents finally said farewell to sharecropping in rural Georgia and used their savings to buy small plots of land in a swampy area, a few miles from the heart of downtown Augusta. For the next twenty-plus years, even as the neighborhood swelled to two hundred families, Hyde Park residents made do without indoor plumbing, running water, or gas stoves. Living on swampland also meant that they struggled through numerous floods, often having to canoe out of the neighborhood to get to work and school. Back then, Hyde Park residents were nearly all employed—many in nearby factories or as domestics. The neighborhood was a vibrant place, with small groceries, churches, barbershops, and even a few bars where residents could take the edge off a long week's work. But for the rest of Augusta, Hyde Park was almost invisible. Surrounded by a junkyard, a railway line, and an industrial ceramics plant on one side, a power plant on another, a brickyard and a second railway on a third, and a highway on the remaining side, Hyde Park essentially formed the hole of an industrial donut. The fact that low-income African American families lived inside that hole made it even less likely to be noticed, especially by Augusta's mostly white politicians.

In the late 1960s, Hyde Park residents, fed up with their lack of county services, caught on to the heels of the national civil rights movement and formed the Hyde and Aragon Park Improvement Committee (HAPIC).[7] After several years of tenacious community agitation, HAPIC activists won water, gas, and sewer lines, streetlights, paved roads, and flood control ditches—infrastructure that most other Augusta neighborhoods had received long before. Throughout the 1970s, as local industries downsized and employment rates declined, they worked hard to improve residents' education, job opportunities, and access to health care. In the 1980s, they added a fight against the selling of drugs and drug-related violence to that agenda, as Hyde Park's streets became a popular local venue for buying and selling crack cocaine.

Some residents also began to fall ill with mysterious or uncommon forms of cancer and skin diseases. Around 1990, they discovered one possible reason for local maladies. In the early part of the 1980s, Southern Wood Piedmont (SWP), a nearby wood-preserving factory, detected soil and groundwater contamination, including dioxins, chlorophenols, and other wood treatment chemicals.[8] On further investigation, SWP found that the groundwater pollution extended off-site along two plumes. The factory closed in 1988 and began to remediate the contamination.[9] Two years later, Virginia Subdivision, a low-income, predominantly white community that backed right onto SWP property settled a class action lawsuit[10] against SWP and its parent companies, International Telephone and Telegraph Corporation (ITT) and ITT Rayonier.[11] Although the ditches that lined Hyde Park's streets eventually made their way to the edges of SWP's property, residents had never been told about the lawsuit, let alone asked to join it; therefore, they received no compensation. Hyde Park residents believed that this exclusion had much to do with the fact that their neighborhood is 99 percent African American.[12] HAPIC geared up for another community-wide struggle, now making environmental justice its main priority. This time, however, rather than fighting Augusta's city hall, the organization took on an international conglomerate.

Hyde Park residents soon realized that Southern Wood was not the only nearby industry to emit toxins. There were reports that Georgia Power's transformer station, which sits on one side of Dan Bowles Road, just across from a number of homes, had leaked polychlorinated biphenyls (PCBs).[13] For years, the plume from Thermal Ceramics, an industrial ceramics plant on the neighborhood's edge, had lit the skies at night. Many mornings, people living just down the road found their cars covered in a fine white dust. Chemicals also flowed from Goldberg Brothers scrap metal yard (about a half a block from Georgia Power) into Hyde Park's ditches and yards. Residents (children in particular) suffered from rashes, lupus, respiratory and circulatory problems, and rare forms of cancer.

An array of studies conducted throughout the 1990s found high levels of certain chemicals in the neighborhood's soil and groundwater, but no conclusive evidence could link those chemicals to health problems. For example, a 1991 study that found higher than normal levels of chromium and arsenic in Hyde Park soil led some experts to suggest that people stop growing vegetable gardens and prevent their children from playing in ditches or dirt around the neighborhood. However, the Georgia Environmental Protection Division (EPD) contended that the levels of heavy metals found in the soil were within normal ranges, and the Agency for Toxic Substances and Disease Registry (ATSDR) determined that community members were probably not being currently exposed to contamination. Yet, Environmental Protection Agency (EPA) tests conducted in 1993 found that the scrap yard on Dan Bowles Road contained widespread lead and PCBs contamination and lesser amounts of arsenic and chromium.[14] Although the yard's owner built an embankment to prevent its water from flowing into Hyde Park, later EPA investigations found high enough levels of PCBs and lead in the yards of at least one home near the junkyard to warrant a $100,000 cleanup. Then, in 2001, an EPA-sponsored environmental investigation of the junkyard led to a $7 million cleanup that removed fifteen thousand tons of waste from the property.

For every study that has shown evidence of significant contamina-

Children playing on carpet laid over dirt, 1999. Photo by author.

tion, there seems to be another that contradicts or mitigates it, leaving residents (and researchers) confused. At least some of the dangerous chemicals just mentioned appeared in nearly every study report,[15] but how intensely they were concentrated, and whether and to what degree they posed a health threat remain very much up for debate. Therefore, residents have won neither a corporate settlement nor government funds to help them clean up or relocate from the neighborhood. At the same time, it became almost impossible for them to sell their homes due to fears of contamination. By the time I came to Hyde Park in 1998 to conduct research with HAPIC, I found that the neighborhood's inhabitants had gotten mired in the (figurative and literal) toxic stew that surrounded them.

Hyde Park residents have thus had the deck stacked against them from the get-go. They grew up under the cruel thumb of Jim Crow and over the years had to fight for neighborhood infrastructure and against the unemployment, drugs, and violence that plagued their community. Then, people in the neighborhood seemed to get sick and to die at an alarming rate. Although high levels of dangerous chemicals had been found in their air, water, and soil, they could not link the two issues according to established scientific standards. What Hyde Park's approximately 250 residents *have* had, however, is a history of successful organizing, as well as a history of being a close-knit, "pull-together" kind of community. They also had a cadre of activists, like Arthur Smith, who managed to hold on to a tremendous amount of faith, energy, and hope, working tirelessly to come up with strategies for improving neighborhood schools, reducing crime and violence, and finding a way out of contamination.

Activist Ethnography

From September 1998 through October 1999, I joined these struggles by volunteering with HAPIC (see appendix A, where I provide a more detailed description of my methodology). HAPIC has never had full-time staff members and relies on whatever after-work time its board members can spare. My work included writing grant proposals, creating a website, organizing after-school and summer programs for youth, and helping plan community meetings and cleanup days. My anthropological background became a useful tool for fulfilling my activist proclivities, and activism, in turn, enriched my anthropological study in many ways. For example, having grown up in a white, middle-class suburb of Washington, D.C., I was a stranger to Hyde Park on many counts. My volunteering for HAPIC made it possible for me to get to know a variety of adults and children in the neighborhood on an informal, everyday basis, as well as providing me with access to information about HAPIC's historical activities. Participating in HAPIC also allowed me to experience firsthand the ups and downs of community activism. Equally important, it enabled me to give something back to these people, who had allowed me to spend time in their neighborhood and to ask them countless questions. Finally, it provided me with a way to exercise my own commitments to activism, which include the writing and publication of this book.

Main Aims

In 2004, black male earnings were 70 percent that of white men, the life expectancy for blacks was six years shorter than that for whites, and blacks who were arrested were three times more likely to be imprisoned than whites who were arrested.[16] Racial segregation, discrimination, and disparities remain pernicious in today's United States, despite the passing of civil rights legislation, the dismantling of ideas about the biological basis of race, and contemporary celebrations of multiculturalism. Why are the life chances of so many African Americans, especially in urban settings, different than those of white Americans? And what on earth does the environment have to do with it?

Somewhere in the middle to late 1980s in the United States, two important social movements—environmentalism and civil rights—converged in an effort to answer these questions. This convergence is taking place around the globe in rural areas, small towns, and big cities, wherever marginalized people realize that they bear the brunt of housing the world's industrial waste. In the United States, environmental justice seeks not just relief from contamination but also access to a host of resources (such as decent housing, schools, and/or police protection) that are traditionally denied to people of color. For this reason, U.S. environmental justice activists initially dubbed their movement "the civil rights [movement] of the new millennium." For many, a civil rights legacy provides a foundation from which activists think about and struggle with environmental problems.

This book shows that environmental racism extends far beyond the straightforward poisoning of air, water, or soil. That Hyde Park residents cannot point a finger at one deliberate polluter is highly typical of environmental justice communities. Their difficulties in meeting established scientific and legal standards to prove their case are also typical, due in large part to the inexactness of environmental science and the biases that it contains. The Hyde Park case is thus emblematic of the complicated and invidious ways in which environmental racism works —how it is embedded in discriminatory institutional practices, policies, and procedures, and how it accumulates over many years. Indeed, one of this book's main aims is to consider the multiple ways in which race and the environment are connected, and how people think about, experience, and organize around such issues in the post–civil rights era. More simply, this ethnography is the story of how one group of activists understood the links between toxic waste and race.

We have few, if any, full-length ethnographies of one environmental justice group. But why devote an entire book to the study of one small organization? My close-up view of one African American group struggling against environmental racism answers questions about how people define and practice environmental justice activism on an everyday basis and in the context of their daily lives and struggles. I show how one group strategized and restrategized as it went through the ups and downs of a more than decade-long environmental justice battle, and more than thirty years of fighting for civil rights. My on-the-ground view thus reveals what inspires people to act collectively to improve their lives, what discourages them from such actions, and how periods that seem to be lacking in social movement activity are actually where the nurturing of the powerful community identities that foster activism takes place.

Apart from being a compelling example of environmental injustice, the case of Hyde Park also encapsulates many of the myriad issues at stake when we talk about environmental justice. For instance, Hyde Park's story shows us how environmental science and environmental laws are skewed toward particular perspectives that do not address the needs of minorities or encompass their experiences. It shows the tenuousness of alliances between environmentalists and grassroots environmental justice activists. Further, it shows how people understand and use the term "environment" in very different ways, which in turn affect how they act toward it. Thus, a detailed, on-the-ground account can refine and sharpen what we already know about political organizing, the pursuit of environmental justice, and who benefits from current environmental policies and practices. Studying one organization in depth also allows me to concentrate on the significance of everyday meanings —how specific cultural histories shape the ways activists explain the terms of their collective actions. For example, Hyde Park residents defined the "environment," "environmental racism," and "environmental justice" according to their specific experiences as southern African Americans living in a low-income urban neighborhood. That subjectivity often contrasts with the ways that white middle-class institutions define such terms.

Today the question of subjectivity confounds activists and academics alike. In this case it presents an obvious paradox: Doesn't activism skew objectivity? Isn't politics the opposite of science? This book shows that all accounts, even scientific ones, are made from particular positions. The Hyde Park position is underrepresented in public discourse; my job as an activist and as an anthropologist is to make it more audible. The risk here is obvious—if even scientists have agendas, then how can scientific facts be trusted? How do we know that environmental damage takes place at all? Does a focus on cultural perceptions of reality only lead us into a relativist labyrinth, where facts are endlessly disputed with other facts? Again, this work itself models an answer,[17] examining who benefits from a given presentation of facts and placing the cultural relativity of experience and perception into a material context. Or, people may understand the environment, race, and poverty in different ways, but certain of those ideas have more power than others. In turn, those dominant ideas have very real effects on the material conditions in which some people live.

For example, pernicious assumptions that poor, urban African Americans are "disorganized," "uneducated," and mired in poverty led to the proliferation and the permanence of unwanted industry and land uses in Hyde Park. It appears that local politicians and corporations assumed residents would not care about, or contest, such sites. As recently as the summer of 2001 (long after Hyde Park's toxic problems had attracted numerous local news reports and even federal grants), Augusta's planning department permitted a local company to establish a landfill in the neighborhood. Even more recently, late in 2003, the planning department approved a new recycling facility directly across from the former junkyard. After twenty years of protest, residents had finally received a federal grant that led to that junkyard's cleanup. Despite all the publicity and effort surrounding that cleanup, someone in the planning department seemed to assume that another landfill or another "recycling center" (often this is a euphemism for "junkyard") would go unnoticed and uncontested in a neighborhood like Hyde Park.

As this ethnography and HAPIC's eventual success in stopping the recycling center attest, the planning department got it wrong. From the inception of their neighborhood and for more than three decades, Hyde Park residents organized against the racial discrimination they faced. Yet, despite the fact that this history of activism can be found in scores of poor black neighborhoods throughout America's cities, we have few ethnographies that depict black activism or the agency of black community members.[18] Thus, this book offers an in-depth documentation of the organization and political mobilization that is characteristic of urban black neighborhoods.

In presenting the results of my ethnographic research on a new kind of activism, I hope that this book will accomplish some activism of its own. I therefore gear it toward multiple audiences, including activists, academics, students, and others. By focusing on the people of Hyde Park and their struggles, I offer an instructive and analytic look at many of the issues that commonly arise as people fight for environmental justice. In so doing, I alert readers not only to the very real, and very complex, problems of environmental racism but also to how one group fought it, day by day. It is my hope that this detailed ethnographic account will provide readers with a deeper and more personalized understanding of environmental injustice, and what can be done to combat it. For those who finish this book and wonder, "What can I do?" appendix B lists contact information for a number of local grassroots environmental justice groups and other resources. Indeed, one of my biggest aspirations is that the stories of the people depicted here—their courage, their stamina, and their consistent ability to adapt their strategies to changing circumstances—will inspire readers to engage in some activism of their own.

Two

Race-ing the Environment

For hundreds of communities of color in the United States, what happened in Hyde Park is not an exception but the rule. Race, numerous studies tell us, is the most potent variable in predicting where hazardous waste facilities are located—more powerful than poverty, land values, or homeownership. Three out of every five African Americans and Hispanics and roughly 50 percent of Asian/Pacific Islanders and Native Americans live in communities containing at least one uncontrolled toxic waste site.[1] The percentage of African Americans or Latinos in a census tract significantly predicts whether that tract hosts a toxic waste facility.[2] African Americans are more than three times more likely than whites to die from asthma, and the hospitalization rate for African Americans with asthma is three times that for whites.[3] Moreover, a *National Law Journal* investigation from 1992 discovered that in minority areas, it took 20 percent longer to put hazardous waste sites on the national priority list than it did in white areas, and penalties under hazardous waste laws were about 500 percent higher at sites having the greatest white population than penalties at sites with the greatest minority population.[4] Such environmental disparities are widespread throughout the United States, but the South has had particularly lax environmental policies. As a result, the region (which houses more than half of the nation's African American citizens) claims eight of the ten states ranked worst in terms of pollution, poor health, and environmental policies. In the EPA's southeastern region, three out of the four largest hazardous waste landfills in the region sit in majority black areas.[5]

This book argues that these startling numbers tell only part of the environmental racism story. Going beyond statistics, I look at how environmental racism affects real people, every day, on the ground, and how those people struggle against this situation, in part by calling attention to the links between environmental pollution and race. In addition, I show how, while race inflects countless aspects of people's lives, living in a poor black neighborhood does not necessarily mean all that popular ideology would have it imply—although such ideas *can* lead to the polluted situations that surround neighborhoods like Hyde Park. Throughout this book, I show just how important definitions are, both for the institutions that use them to assert and maintain privilege and power and for those who work to disrupt such institutions.

Defining the Terms of Debate

"Environmental racism" describes the process that leads to the disproportionate siting of hazardous waste facilities in communities of color. The Reverend Dr. Benjamin F. Chavis Jr., a former executive director of the National Association for the Advancement of Colored People (NAACP), coined the phrase in his foreword to the United Church of Christ Commission on Racial Justice's landmark study, *Toxic Wastes and Race in the United States*. In Chavis's words, environmental racism includes

> racial discrimination in environmental policy-making, enforcement of regulations and laws, the deliberate targeting of communities of color for toxic waste disposal and the siting of polluting industries. It is racial discrimination in the official sanctioning of the life-threatening presence of poisons and pollutants in communities of color. And, it is racial discrimination in the history of excluding people of color from the mainstream environmental groups, decision-making boards, commission, and regulatory bodies.[6]

Many people, however, argue that market dynamics, not racism, are behind the statistics mentioned at the beginning of this chapter. They hold that siting decisions are based on economics, not race—hazards tend to be located on cheap land, and people of color tend to live around cheap housing.[7] Indeed, I do not suggest that every example of environmental racism means some corporation or local agency deliberately decided to put a landfill in a black neighborhood.

At the same time, middle-income black and Hispanic neighborhoods have higher rates of contamination than low-income white neighbor-

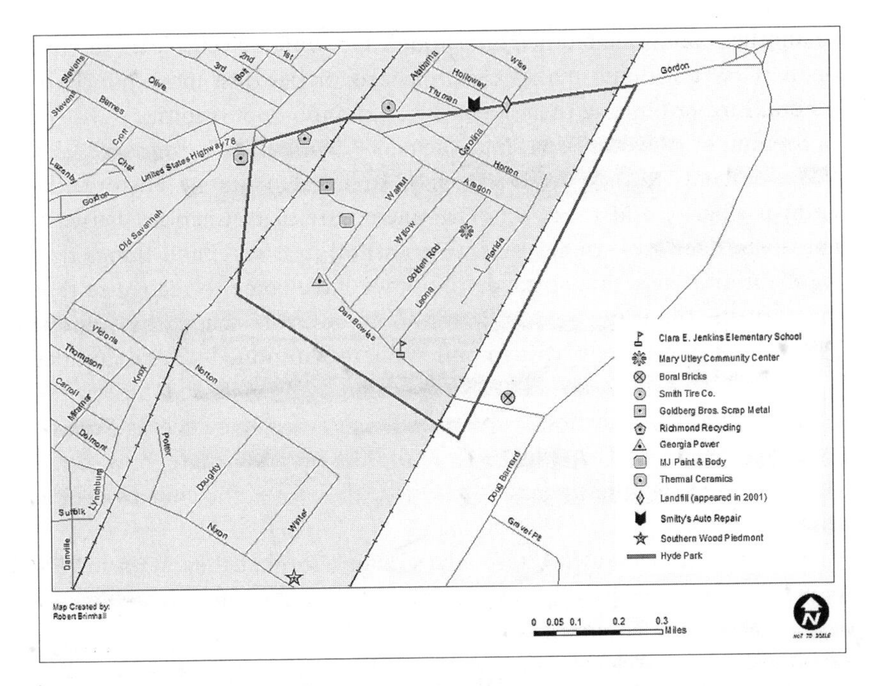

Map of Hyde Park, circa 1999.

hoods, clearly illustrating that the market is not the only influence. The factors that lead to disproportionate pollution and contamination include a host of sometimes subtle racist practices and procedures. For example, real estate agents steer black families toward areas within or near existing ghettos.[8] Minorities with the same resources and credit records as whites are denied mortgages at twice the rate. Minority neighborhoods are more likely than white neighborhoods to be rezoned (often by all-white planning boards) from residential to commercial or light industrial uses.[9] And the list goes on. Due in part to historical exclusions and negative stereotyping, African Americans hold far fewer seats on city councils and other local governing boards. Housing segregation leads to lower tax bases, which lead to poorer schools. Employment discrimination against African Americans continues to prevail in our society.[10] Lower wages and a lack of access to jobs limit the opportunities of African Americans to leave a contaminated neighborhood (although, even if they did leave, they would likely have little choice but to move to another contaminated neighborhood). Growing up with lead, dioxin, or mercury poisoning makes children miss or perform more poorly in school, further limiting their educational and job opportunities.

Seemingly universal concepts such as "wilderness," "nature," and "development" also contribute to environmental inequities. Historically, anthropologists and other scholars made careful distinctions between nature and culture—things that were natural just *were,* and things that were cultural were human-made. Now we have come to recognize that many of the things we see as "natural" are actually shaped by cultural ideas. As cultural theorist Raymond Williams cautioned several decades ago, "The idea of nature is . . . the idea of man in society. Indeed the ideas of *kinds of* societies."[11] Humans create concepts such as "race," "environment," and "pollution," and humans determine how these concepts are used and understood. Often, they serve the ends of people in power.

Scholars call such ideas "discourses" and note that they work in two ways. In the more literal sense, environmental discourse is communication about the environment. From a slightly more theoretical angle, discourse can be defined as a shared set of basic, often unspoken understandings and assumptions embedded in language that gives people a framework from which to make sense of their lives and of the world in which they live. A discourse both "draws on and generates a distinctive way of understanding the world."[12] For Michel Foucault, discourses are a way to assemble bits of disparate information into a coherent whole, and they coordinate the behavior of those who subscribe to them. Although Foucault believes that most discourses are oppressive, I would add that some are liberatory, expressing resistance and alternative worldviews. Moreover, discourses can compete and engage with one another. For example, discourses that view the environment as pristine nature sometimes compete with those that understand it as toxic and discriminatory.

A discursive slant has drawn fire for focusing too much on a rhetorical level and being insufficiently grounded in material life. But deciphering such meanings is far from an idle, academic exercise. As Foucault states,

> It seems to me . . . that the real political task in a society such as ours is to criticize the working of institutes which appear to be both neutral and independent; to criticize them in such a manner that the political

> violence which has always exercised itself obscurely will be unmasked so one can fight them.[13]

A discursive/constructivist approach to the environment, race, or class, therefore, does not mitigate the seriousness or reality of the ecological conditions under which people suffer. The "environment" or "nature" might mean different things for the residents of Hyde Park than for a farmer in Montana, a Native American in Arizona, or an upper-middle-class lawyer in New York City (let alone an indigenous tribesperson in the Amazonian rain forest); but, however construed, the "environment" had very real and lasting effects on the residents of Hyde Park.

Because the causes and effects of environmental racism are so extensive, environmental justice activists define the environment as "where we live, work, and play." For them, the environment is not just ecological but also includes a host of *social* factors such as housing, schools, neighborhood safety, and employment. The pursuit of environmental justice, then, is "the right of all people to share equally in the benefits of a healthy environment." Put slightly less simply,

> Environmental justice initiatives specifically attempt to redress the disproportionate incidence of environmental contamination in communities of the poor and/or communities of color, to secure for those affected the right to live unthreatened by the risks posed by environmental degradation and contamination, and to afford equal access to natural resources that sustain life and culture.[14]

The environmental justice movement encompasses hundreds of activists around the globe, who fight to protect themselves from unsafe environments. Overseas, environmental justice movements include the Chipko, or "tree-hugging" movement against deforestation in India, the struggles of the Ogoni people in Nigeria to stop oil drilling on their farmlands, and the work of indigenous peoples in Malaysia to stop the destruction of their rain forests. In the United States, environmental justice movements range from the struggles of Chicano women in South Central Los Angeles opposing a hazardous waste facility, to Navajo sheepherders fighting to stop the world's largest open-pit coal mine from encroaching on sacred land, to native Alaskans exposed to radioactive waste, to workers in California's Silicon Valley.[15] Environmental injustice affects women, poor people, people of color, and workers, and the issues it raises are global, human rights issues. Although the movement is fairly new, it already has a vast literature.

Over the past several decades, anthropologists have become keenly aware of environmental struggles, as many of our research sites suffer from environmental degradation. Some anthropologists have used the opportunity to publish as a way to depict "urgent case studies" of various environmental justice communities. Most of these studies are contained in edited volumes that describe cases where governments and multinational corporations make deals that remove, poison, or develop the land resources of local and indigenous people around the world. Other anthropologists take a more analytic approach, questioning taken-for-granted assumptions about nature and the environment.[16] Writing a full-length ethnography about environmental justice allows me to do both. First, I present Hyde Park as an "urgent case study" because I believe that it *is* an urgent case. Second, I pay significant attention to the power of language and its role in shaping what we know about, and how we act upon, the environment.

Environmental Exclusions

"The way in which ethnicity enters into the environmental experience," writes historian Robert Gottlieb, "is . . . influenced by a long-standing assumption by minorities that the environmental movement 'belongs' to upper-middle-class or elite Anglo constituencies."[17] The idea that environmental issues reflect elitist interests can be traced to the Progressive Era, which catalyzed reactions against industrial-era progress. Gottlieb summarizes this period as a "response to the urban and industrial changes accelerating with the rapid urbanization, industrialization and closing of the frontier," when Americans began to define the "environment" or "nature" in opposition to the "urban."[18] The acceleration of urbanization inspired the birth of the conservation movement. Led by John Muir, conservationists struggled to find a "cure" for the "weaknesses" derived from city life, which threatened health and virility.[19] According to sociologist Dorceta Taylor, conservation movements embodied "white, bourgeois" values that stemmed from transcendental ideals, which linked the act of communing with nature to transcending everyday worries.[20]

In 1892, the wealthy Muir started the Sierra Club to save the shrinking number of areas where people could enjoy pristine nature. Throughout the 1950s, the Sierra Club remained a "WASP enclave," upholding elitist values and excluding groups such as Jews from its membership. The Sierra Club and the larger conservation movement persuaded the federal government to establish national parks and preservation areas. However, these parks and wildlife preserves that the movement fought for have historically been inaccessible to most minorities due to admission fees, the costs of getting to the site, and/or residence requirements for admission.[21]

In the post–World War II period, a nationwide increase in industrial production and mass consumption resulted in unanticipated wastes and toxic emissions, and the issue of air pollution captured the public imagination. By the 1960s, media outlets increasingly released stories on the harmful health effects of industrial chemicals.[22] In 1964, Rachel Carson published *Silent Spring,* a book that graphically outlined some of the environmental horrors rendered by toxic waste. Her book instantly drew public attention to, and support for, environmental needs, expanding them to include pollution in urban centers.

However, throughout the 1960s and into the 1970s, white middle-class interests continued to frame the ways that urban environmental problems factored into environmental activism, leaving out the greatest number of urban dwellers actually affected by poor environmental quality—minority and low-income communities. In the early 1970s, for example, some epidemiologists and urban activists publicized the fact that an alarming number of inner-city children were being poisoned by lead-based paint. Follow-up reports showed that leaded gasoline also contributed substantially to lead exposure. Within a few years, national environmental groups with existing clean-air priorities took up the cause of reducing lead in gasoline. Because these groups had more resources than urban activists, they propelled the leaded gasoline issue to the forefront of policymakers' agendas, and federal and popular attention to lead paint contamination then faded.[23]

In 1978, the relocation of Love Canal, a small, middle-class community in upstate New York, shocked the nation and turned the course of U.S. environmental activism.[24] Environmental concerns in the United States thereafter included toxic waste and its related health risks, and environmental activists broadened their ideals and goals to cover a wider range of issues, from preserving natural resources to improving clean air and water regulations and opposing nuclear waste. As environmental agendas expanded, environmental movements increasingly institutionalized. Large nonprofits such as the Nature Conservancy, the Sierra Club, and the World Wildlife Fund came to dominate the U.S. environmental political agenda. At the same time, they remained white, middle-class enclaves in terms of membership, leadership, and agendas. Moreover, middle-class opposition to toxic industries frequently called for the closing of factories that employed minority workers. As a result, minority groups often viewed urban environmental activists as a threat to their livelihoods.[25]

Middle-class interests and views of the environment as something to be preserved and conserved have thus shaped U.S. perceptions and policies. In stark contrast, environmental justice activists see the environment as an ecological *and social* resource that is subject to racial discrimination.[26]

Establishing Inclusion

In 1982, an incident in Warren County, North Carolina, instigated an important shift in the course of U.S. environmentalism. In response to being selected for a PCB landfill, a rural, poor, and mostly African American community launched an enormously successful public protest. Joined by such African American political luminaries as Benjamin Chavis (former executive director of the NAACP), Congressman Walter Fauntroy, and the Reverend Joseph Lowery (then head of the Southern Christian Leadership Conference), more than five hundred demonstrators were arrested. The Warren County protest inspired two studies (one concentrating on the South and the other on the nation), both of which concluded that race—not income—was the most significant variable associated with the location of waste facilities.[27] These events fueled the rise of a grassroots, people of color movement. "Environmental justice" became the rallying cry for activists across the country as they connected local health problems to the disproportionate number of polluting sites in their neighborhoods.[28]

Through the late 1980s, the environmental justice movement gained strength. In 1989, the Gulf Coast Tenant's Organization (GCTO), with help from the Southern Organizing Committee (SOC) and the Southern Christian Leadership Conference, organized the Louisiana Toxics March, which included such groups as Greenpeace, the Sierra Club, and the Oil, Chemical and Atomic Workers Union. The United Church of Christ (UCC) established a national committee on environmental racism, and in 1990, EPA administrator William Reilly formed a work group on Environmental Equity. Also in 1990, two social justice groups, the GCTO and the SouthWest Organizing Project (SWOP), initiated a letter to the nation's ten largest environmental organizations criticizing them for excluding minority concerns from their agendas. Environmental justice leaders also began planning a national summit of environmental, civil rights, and community groups.

In October 1991, more than seven hundred people from the United States, Central and Latin America, and Canada appeared in Washington, D.C., to draft a national grassroots environmental justice agenda. Meeting participants adopted the "Seventeen Principles of Environmental Justice," founded on the idea that environmental justice activists (rather than environmental organizations, politicians, corporations, or academics) "speak for themselves." With these principles providing a foundation for the movement, summit attendees decided not to form a national organization but to build locally based movements from "the ground up."[29]

These events, groups, and individuals are frequently listed as those that established today's environmental justice movement. However, other less well-known but equally significant activities occurred at the same time throughout the South, which after the Warren County protests became a maelstrom of environmental justice activity. In 1992, SOC held a meeting in New Orleans that drew twenty-five hundred people. That year, the EPA held a series of regional meetings with its constituents, beginning with the Southeast's Region IV. U.S. environmental justice activists had complained bitterly about Region IV for some time. Its officials, activists contended, would begin cleanup on toxic sites without notifying community members. Thus residents were learning about toxicity in their neighborhoods only after a cleanup was already under way; they had no control over how cleanups took place and were unable to protect themselves from exposures during the process. For instance, in Pensacola, Florida, the EPA sponsored two Superfund cleanups, one of which became the largest in history. Pensacola residents had not yet thought to connect "Mount Dioxin," as they came to call an enormous mound of toxic dirt in their neighborhood, to the thick coatings they found on their cars and windows, or to the high rates of disease in the neighborhood. When their respiratory problems flared just after cleanup began, however, they demanded that Region IV officials take a more collaborative approach. In more widespread instances, the EPA refused to include residential areas in its investigations.

Back in Region IV, on the first day of the 1992 EPA meeting, the *National Law Journal* released its report documenting racial biases in EPA procedures. Environmental justice activists realized that if they adhered to the EPA's existing agenda, there would be little room to voice their concerns. They quickly redirected the meeting's agenda, stipulating that EPA officials hear from impacted community members and not vice versa. Activists also demanded that the EPA clean specific sites. That meeting was the first and last in the regional series, but southern activists sustained their pressure on Region IV administrators, with strategies that included appearing at EPA work sites and staging large-scale protests.

Meanwhile, across the country, regional and local environmental justice leaders traveled door-to-door and "put hundreds of thousands of miles on cars"[30] to alert community members to the dangers of toxic waste facilities. These efforts paid off—at EPA meetings, or at public hearings to address the siting of a new facility, scores of community members turned up to voice their concerns. At the same time, activists maintained their national partnerships. Sustained local pressure, combined with national visibility, eventually led the federal government to take some action. In 1992, the EPA created the Office of Environmental Equity (later changed to the Office of Environmental Justice). A year later, it established the National Environmental Justice Advisory Council (NEJAC), which grew out of one of President Bill Clinton's transition teams. Composed of academics, professionals, and members of impacted communities, NEJAC divided its work among six subcommittees, many of which have been responsible for instigating public policy changes.

Finally, in 1994 President Clinton issued Executive Order 12898, which called for federal agencies and federally sponsored programs to strive for equitable distribution of potentially toxic facilities. The order institutionalized and legitimated environmental justice struggles, but environmental justice activists complain that it "has no teeth" and is not enforceable. Environmental siting decisions are still left up to local and state governmental agencies. Although a community review process is usually required before decisions are finalized, community members are rarely notified of public hearings, which often occur during regular work hours.

Making local governments comply with the executive order is not the only challenge for environmental justice activists. Many have tried to file lawsuits to hold local governments and/or corporations responsible for contamination, but seeking legal remedies from environmental racism is extremely difficult and complicated. Very few lawsuits filed under the equal protection clause of the U.S. Constitution (in other words, specifically claiming environmental *racism*) are successful, and only a handful of plaintiffs filing other kinds of claims (such as Title VI, which regulates siting procedures) have won their cases.[31] Environmental justice activists thus combine legal battles with extralegal strategies (HAPIC activists, for example, both filed a lawsuit against Southern Wood Piedmont and appealed to state and federal environmental protection agencies to fund the cleanup of their neighborhood or to relocate residents). Even these appeals are rarely successful, however, as environmental cleanups are extremely costly, and relocation is even more expensive.[32]

Many tactics used in the environmental justice movement derive from the civil rights movement,[33] which is not surprising, since many of its participants are veteran civil rights leaders. Consider the following two examples.

Like many southern African Americans of his generation, Gary Grant's activism began when he was growing up (in Grant's case, in Halifax County, a rural part of North Carolina). His parents challenged voting laws in the 1950s, and they signed him up in an unsuccessful effort to integrate a local high school when he was in his early teens. Grant also went to an interracial camp during his teen years, which one summer was targeted for a cross burning by the Ku Klux Klan. Several years later, Grant left his rural home for North Carolina College in Durham. There he stepped up his activism, getting arrested "many times" for his involvement in sit-ins at Howard Johnson's and other local institutions. Grant was the first African American to sit downstairs at the Carolina Theater in Durham, which today is a historic marker in North Carolina's civil rights history.[34]

Returning to Halifax County as a teacher in 1965, Grant integrated his local Boy Scout Jamboree and then led a group to protest the closing of the all-black school that employed him. That group soon began to explore black land loss issues. In the 1990s, this effort spun off into the much larger black farmers' movement, which filed a highly publicized and successful lawsuit against the U.S. Department of Agriculture demanding reparations for the loss of black-owned farms. Also in the 1990s, Grant and his neighbors were notified that their community was targeted for between fourteen and seventeen new industrial hog operations. They began connecting black land loss issues to environmental racism—the fact that they had lost control of their land meant that they had also lost control of regulating their environment. Grant also discovered that the average age of male mortality in his community was fifty-five. Immediately, he launched an effort to oppose the influx of the hog industry. Over the next few years, his organization successfully lobbied for the passing of state legislation that limited livestock operations.

In a second example, Doris Bradshaw had been an activist "just about ever since [she could] remember." Her father became president of the NAACP in Alamo, Tennessee, in 1964 and eventually had the local sheriff arrested for beating two black high school–aged boys. Bradshaw was also one of the first eight African Americans to integrate Alamo High School, a process she described as "horrible." After her intense experiences as a youth, Bradshaw moved to Memphis and decided that she "wanted out of everything," spending the 1970s and 1980s raising her family—that is, until 1994, when she got involved in her children's Parent-Teacher Association. Around that time, she noticed that she was attending funerals almost every weekend, and her elderly neighbors seemed to die of strange illnesses. Bradshaw connected these deaths to the fact that her neighborhood was located adjacent to the Memphis Defense Depot. Once the largest military supply storehouse in the country, in 1992 the depot was placed on the Superfund National Priority List because a wide array of contaminants, including arsenic, cadmium, chromium, lead, mercury, pesticides, dioxin, PCBs, and chemical weapons residues were found on-site.

Bradshaw soon revived her activism, leading a fight both to hold the Defense Department responsible for community health problems due to contamination and to secure in her community the same things she fought for as a teenager—equitable education, employment, and treatment in general for all people, regardless of the color of their skin.

Dig a Little Deeper: Race, Class, and the City

If environmental justice is "the civil rights of the new millennium," then what happened to the civil rights of the old millennium? In 2004, the National Urban League reported that homeownership and college enrollment rates among blacks were at an all-time high. Black unemployment, however, was more than twice as high as that of whites, and black poverty rates were three times higher than those of whites. Startling disparities also existed in the quality of education that blacks and whites receive, where they live and how many own homes, and their arrest and imprisonment rates.[35] In addition, African Americans are twice as likely to die from disease, accidents, and homicide than their white counterparts. A 2002 study by the National Institute of Medicine found that racial minorities receive lower-quality health care than white patients, even when their income, insurance coverage, and medical conditions are similar.[36] A 2001 Gallup poll reports that only one in ten black Americans, regardless of their income level, believes that blacks are treated the same as whites in the United States. These examples powerfully suggest that economic and social playing fields for blacks and whites have not leveled, even in the wake of post–civil rights legislation.

By studying environmental justice activism in a poor black neighborhood, this book aims to show how these national statistics are both lived and contested. For instance, much of the organizing I encountered challenged popular assumptions about the connections between race, class, and the city. The danger of these assumptions is that they shape what many people know about black urban life and have "strongly influenced, if not determined, the terms of political debates in the United States concerning race, social equality, and the changing political economy of American cities."[37] Viewing poor black neighborhoods as isolated, apathetic enclaves of violence and crime engenders public policies that dismiss the voices, opinions, and even rights of individuals living there. In Hyde Park, such stereotypes led to the proliferation of toxic waste sites. Local politicians and corporations assumed that such a "disorganized," "uneducated" community, mired in poverty, would not care about or contest such sites.

As they have studied urban poverty and its racial implications, social scientists, unfortunately, have reinforced some of these stereotypes.[38]

Early on, a number of anthropological studies in the United States and elsewhere contributed to notions about the biological basis of race.[39] Then, in the late 1930s, American anthropologist Franz Boas took a bold step away from such views. Proving that vast physical differences existed within racial categories, Boas contended that there could be no scientific basis for biological differences between races.[40] Anthropologist Ashley Montague joined Boas when he published *Man's Most Dangerous Myth: The Fallacy of Race,* which painstakingly challenged prevailing ideas about race and biology, pointing out that there are no lines in nature that distinguish separate races.

During the 1930s and 1940s, while most anthropologists were studying "exotic" societies, a small number followed Boas's lead and studied racism and race ranking in the southern and northeastern United States. These books, especially Gunnar Myrdal's well-known *An American Dilemma: The Negro Problem and Modern Democracy* (1944), pointed out how U.S. egalitarian ideals contradicted the practice of racial segregation and oppression.[41] Taking a slightly different tack, sociologist E. Franklin Frazier studied black families in a Chicago "ghetto." Rather than focusing on race relations or race ranking, Frazier looked at black families and hypothesized that a history of slavery had left them "disorganized" and "without aims and ambitions." With their African culture taken from them, and after so many years of suffering, African Americans simply could not organize themselves enough to emerge from poverty. Building on this concept, in the early 1960s anthropologist Oscar Lewis argued that having been deprived of economic resources for so long meant that poverty had become a way of life for many people. To adapt to their economic conditions and to cope with "feelings of hopelessness and despair which develop from the realization of the impossibility of achieving success in terms of the values and goals of the larger society," the poor had developed a list of pathological behaviors.[42] These behaviors were then reproduced from generation to generation, and they prevented impoverished people from organizing for political change.

Both Frazier's and Lewis's ideas, which characterized poor African Americans as fatalistic, disorganized, and resigned, reflected common liberal assumptions about the helplessness and passivity of the poor, who could not break cycles of poverty on their own. They also explained why African Americans apparently did not revolt, protest, or move ahead. Such ideas then coincided with a report by Daniel Patrick

Moynihan (Lyndon Johnson's secretary of labor), alerting the public to a new, urgent crisis in cities: family breakdown created a "tangle of pathology" in which women were heading households and black men had all but dropped out of sight. The Moynihan report generated a flurry of scholarly and popular attention to poor urban African Americans (who would later become known as the "underclass").[43] Like culture of poverty theses, most of these studies conveniently blamed the problems of the urban poor on the poor themselves—they had an improper family structure, they had sex too early, they hustled, they were loud.

Black scholars and activists raised a great hue and cry in response to such ideas, and a number of anthropologists joined in to say "not so." The opportunities that many Americans take for granted, they contended, are blocked to poor people of color, who continue to hold mainstream values *despite* the fact that economic oppression and institutional discrimination prevent them from achieving the American Dream. For example, Swedish anthropologist Ulf Hannerz asserted that "ghetto" culture consists of mainstream ideas as well as specific adaptations and reactions to poverty.[44] The idea that poor people of color engage in adaptive and creative solutions to their conditions soon became quite popular among anthropologists. Carol Stack's *All Our Kin,* one of the best-known examples, argued that poor black families strategically adapted to poverty through extensive kin and friend exchange networks. Similarly, Joyce Ladner found that adolescent black girls adjusted to life in an impoverished inner-city neighborhood in creative and healthy ways.[45]

While these important studies debunked many ideas about poor people's behavior, they also reinforced the idea that urban blacks live in isolated enclaves. Such notions were further popularized in the 1980s, when income levels in the United States began to polarize even further. Skyrocketing inflation in the 1970s, combined with that decade's energy crisis, led to dramatic welfare cutbacks. The recession that ensued, along with Reagan-Bush "trickle-down" policies, further split the rich and poor, until the United States was left with the dubious honor of having the first world's largest number of poor people and smallest middle class, proportionately. Responding to these distressing disparities, the 1980s produced a flurry of "underclass" articles and books, most of which reinvented the culture of poverty thesis and rationalized the era's domestic policy.[46]

For example, journalist Ken Auletta's *Underclass* (1982) described a permanent minority among the poor who were passive, hostile, and/or traumatized. Several years later, Charles Murray's *Losing Ground* maintained that the extension of welfare programs in the 1960s caused shiftlessness and sloth. Welfare dependents, he argued, work the system by having more babies and refusing employment. Taking a slightly different tack, political scientist Lawrence Mead posited that welfare creates a population unable to work, vote, or function as U.S. citizens. Journalist Nicholas Lemann blamed sharecroppers for bringing an underclass lifestyle (characterized by crime, poor education, and teen pregnancies) with them to northern cities.[47] More recently, we have heard that African American youths hold one another back by disparaging each other for performing well in school.[48] In these views, *all* ethnic and racial groups have similar life chances; only the maladaptive behaviors of *certain* groups keep them from attaining upward mobility.

A number of social scientists have pointed out the fallacies of "underclass" depictions. For instance, contrary to popular belief, most poor Americans are white, and most people on welfare are not African American. Moreover, federal government studies show that black adolescents consume illegal drugs at lower rates than whites, black and white pregnant women consume illegal substances that may injure their fetuses at the same rates, but doctors report black women to law enforcement authorities ten times more often, and federal data show that the higher his income, the *less* likely a man is to make child support payments. Finally, values like fortitude, hard work, and honesty are of primary importance in *every* community.[49] Moreover, theories that "blame the victim" fail to take into account the structural causes of urban poverty. In *The Declining Significance of Race* (1978) sociologist William Julius Wilson argues that increasing technology and automation, industry relocation, and labor-market subordination reduce the life chances of unskilled African Americans either by pushing them out of the workforce altogether or by relegating them to low-wage, dead-end jobs. Wilson thus shifts the focus of the underclass debate from blame-the-victim rhetoric to changing economic processes.

In many ways, however, Wilson also reinforced negative stereotypes and misperceptions of poor black neighborhoods. For example, he proposed that "black life chances [have become] increasingly a consequence of class affiliation,"[50] and thus discrimination no longer prevents African Americans from moving into better jobs or neighborhoods. That

African Americans make up the majority of the American "underclass," in Wilson's view, had far more to do with social isolation than with contemporary racism.

Wilson's ideas coincided with a growing movement among liberal academics (particularly anthropologists) to reinforce the thesis that race has no biological basis and that it is socially defined.[51] In their zeal to debunk connections between biology and race, however, many anthropologists characterized race as simply an extreme case of ethnicity. Race even became a "divisive" and "impolite" topic.[52] This emphasis strengthened the ideas of those neoconservative thinkers who blamed certain groups of people (e.g., African Americans) for remaining on a low rung of the socioeconomic ladder. If race does not really exist, and if we can classify all social differentiation under the umbrella of ethnicity, then African Americans ought to have been able to assimilate into mainstream culture just as well as any other group.

These evasions were amplified as Wilson expanded upon his economic theses in *The Truly Disadvantaged* (1987). Here Wilson proposes that underclass neighborhoods stay that way, in large part, due to middle-class desertion. In the old days, he writes, middle- and lower-income blacks lived together, helping each other out and serving as positive examples. But, after the passage of civil rights legislation, the black middle class took advantage of increased educational and occupational opportunity and disappeared from urban areas, leaving those areas filled with unemployed black men and unwed women who do not share the community values of bygone days.[53] This notion was soon echoed in popular laments about the absence of "old heads" and role models from "ghetto" areas.

I contend that these ideas are problematic for several reasons. First, upwardly mobile blacks have always moved up and out of the communities they came from. In fact, Franklin Frazier, and St. Clair Drake and Horace Cayton documented residential mobility in the early part of the century.[54] Second, Wilson's theory wrongly bifurcates black class structure—either you are an impoverished welfare recipient who lives in a neighborhood plagued by crime and violence, or you are a middle-class opportunist who has turned your back on your roots.[55] Hyde Park shows us that this binary is far too neat.

On the one hand, Hyde Park residents themselves frequently described their neighborhood as a "ghetto." Indeed, unemployment and crime rates (especially drug selling) were high, and many residents were poor. In 1998, 47 percent of Hyde Park residents earned less than $10,000 per year, the median household income in the neighborhood was $8,983, and 67.4 percent of residents lived below the poverty level.[56] Many people struggled every day to pay their light bills, make their car payments, and afford their rent. A good number lived without air-conditioning, in homes that badly needed repairs and were overcrowded with kin. Moreover, more than 50 percent of residents were over fifty, lived on fixed incomes (either Social Security, Aid to Families with Dependent Children, or Temporary Assistance for Needy Families), and had spent at least twenty-one years in their current home. On the other hand, income levels *did* vary. Slightly more than one-fifth of residents reported incomes of $20,000 per year or more.[57] The neighborhood was home to teachers, ministers, bricklayers, pipe fitters, secretaries, social workers, and cooks, as well as families who received federal aid. Thus, I found much economic diversity and many gradations of family income within this "ghetto" area—Wilson left out all those people who worked one, two, or even three low-paying jobs that barely kept them afloat.[58]

Perhaps most important, many Hyde Park residents who had managed to earn enough money to move out of the neighborhood had not turned their backs on it. They returned often to help their families and friends in a range of ways, from organizing for HAPIC, to fixing a front porch, to minding a store. Some also returned to attend Christian Fountain, Hyde Park's main church, or simply to visit CE's Place, a neighborhood bar. Hyde Park residents were by no means isolated or lacking in role models or community networks. Rather, activism is alive and well among both present and former Hyde Park residents. The neighborhood is close-knit, organized, and highly functional. Neither their behavior nor their relationships kept people impoverished. Period.

In fact, as sociologist Mary Pattillo-McCoy points out, having a higher income does not necessarily mean that a black family will escape the effects of discrimination. Rather, a neighborhood's racial makeup frequently determines the quality of its schools, its security, its political clout, the appreciation of property values, and the availability of desirable amenities.[59] Indeed, as the lives of Hyde Park residents attest, "Race has become all too real in its social ordering of perceptions and policies. . . . For worse, not better, today we all live in a racialized world."[60] Race may not be biological, and it may be mean different things at different points in time, but it has by no means receded in social import or become a figment of our collective imagination or a remnant of days gone by.

At the same time, while racial identities often acted as a prism through which people in Hyde Park viewed their lives, it would be unfair to say that they *always* viewed things through a race-identified lens (or, for that matter, that all Hyde Park residents had uniform perspectives or practices). For instance, one evening Michelle, a Hyde Park resident, told me about some problems she was having with her white boss. I asked whether she thought it was a "black-white" thing. "I thought that at first," she replied, "but I think it's just him." Michelle's experience demonstrates how race was a primary but not overdetermining factor in how residents viewed their social relationships. Indeed, the experience of African Americans certainly does not dictate that they act or think a certain way. But these experiences (especially segregation and discrimination) have had a particular impact on how social and cultural life is organized and enacted. As one middle-class activist opined, "Being black biases you to think in racial terms."

My goal as an ethnographer is to develop an understanding of the deep significance that certain concepts, such as race, have for people's everyday lives. I am especially interested in how race mobilizes grassroots political action. For although there is certainly no such thing as *a* black community, the notion carries great weight among whites as well as blacks: appeals to racial solidarity work.[61] In investigating that phenomenon, this book travels down a surprisingly seldom-trodden path. Although many have studied urban black life, we have few in-depth accounts of contemporary urban black activism, especially on the level of grassroots organizing. A second goal of this ethnography, then, is to counteract that imbalance and show how neighborhoods like Hyde Park have long histories of sustained activism.

Explaining Activism Ethnographically

W. E. B. Du Bois once wrote that after Reconstruction, "Democracy died save in the hearts of black folk."[62] But democracy has continued to live in the hearts of many Americans, and many of those people fight every day to make democratic ideals match their experiences and to reimagine and remake the world. As it examines black activism, this book seeks new answers to old questions about why and how people mobilize to change the circumstances of their lives. Because I study social movements by observing and participating in them, I answer questions about what democracy means and how it is achieved, through activists' own perspectives. That is, I examine what activists do, as well as what they say about what they do and why they do it.

Indeed, how can we understand the impact of social movements on people's lives if we do not understand how people themselves define "success"? What fortifies or weakens a movement? Why do some movements last and others disband? How do factors such as shared emotions,[63] democratic values, and experiences sustain a movement and keep it on course, even during seemingly latent periods? What strategies do activists draw upon as they struggle to alter everyday meanings as well as their material conditions? This on-the-ground account answers such questions by investigating how social action is understood, enacted, and lived day by day.

The everyday experiences of HAPIC activists add much-needed complexities and refinements to existing theories of social movement organizing.[64] After the sweeping upheavals of the 1960s, a number of social scientists reinvigorated the study of collective action, but they did so mainly by developing far-reaching and overarching models of collective behavior. In the decades that followed, this work fell roughly into two camps.[65] In the United States, the "resource mobilization" (RM) approach frequently explained civil rights movements. In Europe, the "new social movement" (NSM) approach often drew upon environmental movements to support its perspective on social action. In my description of the convergence of environmentalism and civil rights, the RM and NSM approaches also converge.[66]

In very general terms, RM, or "strategy-oriented," theorists look for the structural causes of mass action and the resources (including material, human, and organizational) available to social movement actors. This approach concentrates on the goals of a movement, the means used to implement those goals, and movement outcomes. For instance, many RM scholars examine the kinds of resources available to African Americans in the 1950s and 1960s (e.g., support from the church, from middle-class blacks, and/or from liberal white northerners) and the different networks through which oppositional consciousness was mobilized (e.g., church organizations, historically black colleges, face-to-face interaction). Some also analyze why civil rights movements declined in the 1970s and evaluate whether or not they achieved specific policy objectives. In its focus on material objectives and outcomes, however, the RM approach has been taken to task for its narrow definition of "success," which, for example, excludes objectives such as cultural transformation. More important for this study, it also excludes activists' own ways of defining the outcomes of their movement—evaluations that reflect both concrete, observable results and less tangible ideas about success. Nor do RM theorists address movements that mobilized on the basis of very few resources, the movement's internal dynamics, or some of the less obvious forms of protest that might also constitute collective action.[67]

The NSM, or "identity-oriented," school of thought derives from Europe and focuses on less obvious forms of protest, proposing that social movement actors construct their movements largely through symbolic practices. For NSM proponents, post-1960s movements are "new" because they mobilize around cultural and symbolic issues rather than strictly economic or political issues. In contrast to movements prior to the 1960s, which focused on reforming political and economic policies, NSM scholars claim that recent forms of feminism, environmentalism, and assorted other movements strive "to change cultural and social assumptions." Rather than seeking to overthrow existing political or economic structures, new social movements work on reconstructing public opinions, values, and forms of behavior.[68] In the NSM paradigm, participating in a movement is often an end in itself.

But how does this approach accommodate activists who have material *and* symbolic goals in mind? And are these movements really *new,* or do some of them extend older, more traditional movements? Throughout this book, I describe how as they attempted to counter and confront stereotypes about themselves and the meaning of the environment, HAPIC activists also worked to change their ecological and economic circumstances. In addition, their experiences demonstrate that a legacy of black activism provided not only a foundation but also a continuing thread that was carried from the past into the present. In other words, I found that what is often construed as "new" was "old," and vice versa.[69]

Moreover, the time I spent attending HAPIC meetings and events and listening to the organization's members allowed me to glimpse some of the ways that activist consciousness is nurtured over time. Social movement scholars frequently emphasize organized protests and events and overlook periods of inactivity. This oversight obscures the ways in which social movements remain latent and shore up resources. Sociologist Verta Taylor, for instance, finds that after 1945, certain organizations cultivated and sustained women's rights activists, holding the women's movement in abeyance until it resurged in the 1960s.[70] HAPIC served a similar purpose. Most Hyde Park residents had lived in the neighborhood for at least half, if not all, of their lives. HAPIC provided a channel through which they could build narratives emphasizing community solidarity against various forces (including local politicians, floods, or corporate pollution). A history of conflict coupled with a venue for discussing and narrating those conflicts fostered a strong and specific neighborhood identity, which in turn fed social activism in the present.

Paying attention to the subtleties revealed through day-to-day action also reveals that both strategy and identity (i.e., material and symbolic efforts) are critical to social movement action; indeed, they are two sides of the same coin.[71] At times, HAPIC members characterized themselves as environmental victims because of their race. At other times, they emphasized their unity with all people, reasoning that pollution is color-blind and affects all races alike. In this context, each presentation of identity (one that called attention to racial distinction and one that blurred it) was aimed at gaining the support of a particular audience. Thus, the ways in which activists talked about and communicated their group identity shifted according to the strategic needs of a given situation, making the articulation of identity itself a movement strategy.[72] By the same token, certain strategies required an emphasis on a particular identity.

My ethnographic study of social movement organizing thus suggests how, given the opportunity to observe and participate in the daily goings-on of social movement organizations, we find what should already be obvious: political struggle is not linear or neat. Social movements are made up of real people, who are in the process of reimagining the world. They can be contradictory and emotional, as well as rational and strategic. Just as there is no single black community (despite what social scientists and the media tell us), just as the meaning of the "environment" is not necessarily confined to ecological resources, and just as racism is not solely contained in the beliefs or practices of individual actors, social movement dynamics are overlapping and unruly. They defy boundaries, models, and temporal isolates.[73] It is this messiness that this study demonstrates most of all.

Three

Old Heads

In 1946, Peter Saulsberry, a tall and still powerfully built octogenarian, paid fifty dollars for some land just outside of downtown Augusta. His fifty-by-one-hundred-foot plot lay smack in the middle of a large triangular field of swampland stretching between two railroad tracks and the Merry Brothers Brickyard. Across another field, the Piedmont Wood Preserving Company processed telephone poles.

Peter grew up just outside of Waynesboro, Georgia, where his parents were sharecroppers. In the early 1940s, he was drafted into World War II and became an officers' barber. Once he was released from the service, Peter moved to Augusta and began doing carpentry work. His brother-in-law had just bought two plots of land in a new African American development, and Peter followed suit. Soon he moved into his first house.

Back then, Peter's only neighbors were a couple who sold barbeque out of their one-room cottage. But it did not take long until he was surrounded by relatives, including a new wife, his brother-in-law, and a nephew. Fellow sharecroppers from Waynesboro also began to fill the swamp with homes, and the neighborhood started to thrive. Peter added a barbershop to the back of his house and enjoyed a brisk Saturday business in addition to his carpentry work—in fact, he built many of Hyde Park's houses with his own hands.

In 2003, those hands no longer built much, but they still barbered. Although his wife had passed away some twenty years earlier, Peter was far from lonely or idle. He was an active grandfather and a deacon at his church. He also maintained close kin relations with his neighbors. "I have a wonderful neighbor there," he told me, pointing one long finger toward his window, "a wonderful neighbor across the street. I can't say about those on this side, but these three houses I have wonderful neighbors. We look out for each other."

Deacon Saulsberry, 1999. Photo by author.

Indeed, on any given Saturday, a few "old heads" and their offspring would drop in for a haircut. Tilting his gray head and stroking his moustache, Peter would proclaim that he was Hyde Park's second and oldest living resident, and then fill their ears with news of the present and tales of the past.

Totsie Walker lived alone in a small, brightly whitewashed house hidden behind a lattice fence. In May 2003, we sat in her living room for a chat. Totsie, who was nearly swallowed up by her large, faded armchair, had been receiving guests in that same room for more than forty-one years. But ever since the "high water" of 1990, Totsie believed that her living room had been sinking. Not that she could afford either to repair it or to move. "If I tried to sell [my house], I wouldn't get nothing for it," she said, shaking her head.

Like most of Hyde Park's early residents, Totsie was born to sharecropping parents in Waynesboro, Georgia. In 1950, her new husband, John, persuaded her to move to the city of Augusta, where they could find better employment opportunities. John found work first for the local gas company and then driving trucks for the highway department. Totsie cleaned houses for families living "on the Hill." Finally, they saved enough money to buy property of their own in Hyde Park, where Mr. Walker's first cousin already lived. Totsie remembered her reluctance to move to "the Park":

> *I didn't like it because I heard someone saying something about water was out here. I fought to keep from coming out here. I prayed about it and whatnot, and something said, "Well, go with him and maybe after you get there you can move somewhere else," but I got stuck here.*

Once they were settled, Totsie and John began trying to have a family of their own. "I always wanted kids," Totsie told me, "but God didn't see for me to have kids." When John's sister was unable to keep her son, the Walkers adopted him. Eventually, Totsie grew used to life in Hyde Park and, like Peter Saulsberry, found that her immediate family may have been small, but her kin network was large. "Well, I got to know my neighbors after I moved in here. My neighbors on this side was just like family," Totsie said. Indeed, she spent a lot of time helping family members and friends. "I enjoy going around and helping out. Long as I was able, that's all I did all my life."

As Totsie neared the end of her fifties, however, her life took a tragic turn. Her adopted son, who had gotten a job at Southern Wood Piedmont, died suddenly of heart trouble. "He was working one day and went in to take a shower and fell dead," Totsie said. She could not say for sure that there was a connection between the contamination found at Southern Wood and her son's untimely death, but she certainly did not rule it out. Then John fell ill with diabetes and heart trouble, and she had to cut back her work hours to take care of him. After a few years, Totsie quit working altogether and nursed John until he passed away from a stroke in 2000. By that time, Totsie felt she was too old to go back to work. A year or so after John's death, she suffered a mild stroke. Although she recovered pretty well, in 2003, Totsie continued to have trouble with her right leg. She also took pills to improve her circulation and lower her blood pressure.

These medications were expensive, and their cost had a lot to do with Totsie's feeling stuck in Hyde Park. Her Social Security checks amounted to approximately $620 per month. The circulation pills cost $99 per month, and the blood pressure pills cost $200 per month. Medicare covered none of them. "By the time I pay for the medicine, it's a good thing I don't eat much," she sighed, with a hint of sarcasm.

Totsie's long life was similar to those of many of her neighbors in that it was marked by poverty and sometimes inexplicable family illnesses. At the same time, Hyde Park neighbors treated one another like kin, and many had known one another for most of their lives. On pleasant days, her neighbors could see Totsie leaning on her "stick" and heading up the road to the Mary Utley Center. There she would visit with old friends and, like Peter, regale them with tales of Hyde Park, past and present.

In between the Tracks

The golfer Tiger Woods made history in Augusta, Georgia, in April 1997, when he became the first person of color to win the Masters Tournament. His record-breaking win also entitled Woods to become the third member of color in Augusta National Golf Club,[1] which until 1991 prohibited blacks. The Masters Tournament and Augusta National are Augusta's main claim to fame, and the city's economy relies on the tournament's popularity. In 1999, for instance, the Masters brought $100 million into the city.[2] Unsurprisingly, three years before Woods's win, Hyde Park activists chose Masters week to protest the contamination of their neighborhood. Hyde Park activists not only wanted to win the attention of the approximately 250,000 corporate bigwigs and wealthy golf enthusiasts who annually vie for space on Augusta National's pristine greens but also wished to highlight the striking racial disparities in their city. For approximately seven miles away from Augusta National's rolling hills and immaculately kept lawns, Hyde Park sits seething with toxic chemicals.

The striking disparities between Augusta National Golf Club and Hyde Park's physical landscapes, and of the role of the Masters Tournament in the lives of various Augustans, symbolize the race and class disparities that make up Augusta's social environment. If you ask a rich, white Augustan to tell you about the Masters, for example, he or she will likely describe the club's beautiful rolling lawns and the parties that accompany the tournament. Ask a white teacher or a college student about it, and he or she will probably describe the various opportunities that the Masters presents for making extra money. Ask a black activist like Terence Dicks, and he will make a sly reference to the tournament's title and Augusta's slaveholding history. Finally, ask a Hyde Park resident about the Masters, and that person will probably recall the year that he or she staged marches and protests throughout the tournament.

This chapter outlines how Augusta's uneven social environment developed over time, particularly in terms of its race relations. Just as the environment can be conceived of in many ways, it can also be contaminated in many ways. Augusta's environment is poisoned by historical racial biases and inequalities.[3] The racial imbalances that have guided the city's political and economic history have led to the social and ecological contamination of its African American neighborhoods. Put more simply, this chapter sets a context for HAPIC's current activism by showing how the past poisons the present.

That process works in two ways. First, historical structures, practices, and beliefs persist into contemporary life. For instance, due to the historical disenfranchisement and disempowerment of blacks, very few held seats on the city council until the late 1990s. This imbalance, in turn, accounts for why it took a major battle for Hyde Park residents to receive the infrastructure (i.e., running water, paved roads, and gas lines) that most other Augustans already had. In 1996, the city and county governments consolidated, and blacks established a much greater presence in Augusta politics. But almost every hotly debated issue that came before the county council—no matter how seemingly race-neutral—ended up pitting black council members against white members. As a result, moving forward on city business required much deal brokering. The needs of neighborhoods like Hyde Park can easily get left out of such negotiations—with sometimes drastic consequences for neighborhood residents. Second, history reveals that race relations are not natural. No one knows this better than African Americans in the South, who have persistently fought for better treatment and equal rights. Outlining a history of race relations in Augusta, and how black Augustans continually contested those relations, highlights how people do not forget past injuries such as slavery, segregation, and discrimination. These injuries underlie present understandings, ideas, and practices.

A Big, Small Southern Town

> [Savannahans] have a saying: If you go to Atlanta, the first question people ask you is "What's your business?" In Macon they ask, "Where do you go to church?" In Augusta they ask your grandmother's maiden name.[4]

Augustans often say that they live in a "big, small" town. What they mean is that with nearly two hundred thousand people, Augusta–Richmond County is considered Georgia's second most populous metro area. At the same time, it retains much of its rural character and frequently feels like a small town. Like many midsize southern cities, Augusta originated as a "port city," an urban center in the midst of a largely rural region, which exploited raw materials from those rural areas and imported finished products from more industrialized, cosmopolitan centers. As a link between agrarian and industrialized regions, Augustans maintained a rural ethos, even as their city grew.[5] Because the South remained agrarian longer than the Northeast, such hybridity is fairly common in the region, as are "big, small towns" and "old boy networks." On a social level, that particular mixture means that power generally stays in the hands of a core cadre of people (both white and black), and if you want to count degrees of separation between residents, three is a more likely number than six.

Being caught between large and small also means that Augustans suffer from "stepchild" syndrome.[6] For example, Augusta is second to Atlanta in terms of population, but it is a far distant second. With a population of 3.3 million people, Atlanta has Augusta pretty well beat in terms of new industries and promises of future prosperity.[7] Augusta is also not nearly as well known or well visited a city as Savannah or Athens, which means Augusta receives state-level infrastructure improvements long after other cities do.[8] Additionally, Augusta is a relatively poor city. In 1999, 30 percent of Augusta–Richmond County's families were ranked "low income," 19.6 percent of people lived below the poverty level (compared with 13 percent in the state overall), and the unemployment rate between 1996 and 2000 averaged 6.8 percent (compared with the state's average of 4.2 percent)—which does not help its statewide clout.[9] A feeling of getting short shrift thus runs through Augustans' interpretations of politics, both past and present, and plays a significant role in shaping the city's unique social relations.

Indeed, Augusta both is and *is not* like other southern cities. Urban historians have tended either to bypass southern cities altogether or to focus on the question of their uniqueness relative to northern cities. My focus is on Augusta's uniqueness relative to other cities *within* the region. Both in the American imaginary and in American anthropology, the South is often essentialized.[10] Yet significant political, economic, and social differences exist between and within southern states and regions. As southern anthropologist Carole Hill notes, some macro processes (such as globalization, economic booms, and recessions) have affected certain areas of the South but left others virtually untouched.[11] For instance, while metropolitan areas such as Atlanta, Dallas, and Charlotte have benefited tremendously from post–World War II economic growth, others, especially rural areas, have been unable to replace revenue losses from declines in agriculture-related industries. Semimetropolitan areas, such as Augusta, fall somewhere in between. But even though in their peculiar mix of urbanism and ruralism they offer important insights into historical and contemporary southern life, midsize southern cities like Augusta *are* generally treated as "stepchildren" by social scientists.[12]

At the same time, although "*the* South" might not exist on a practical level, on metaphoric and symbolic levels, it holds substantial meaning, especially for those who live or grew up there. In his book *The Mind of the South* (1941), W. J. Cash famously argued that a common, southern ethos distinguished the region from other places in the United States through the institution of slavery. By owning slaves, wealthy southerners saw themselves as members of a "planter class" akin to nobility. A social hierarchy followed from that, ranking poor whites significantly higher than blacks. For Cash, the desire to maintain that hierarchy, along with economic benefits derived from a slave system, inspired white southerners to declare war. As southerners subsequently faced myriad hardships during the Civil War and then Reconstruction, a collective southern identity and solidarity was cemented.[13]

Cash's arguments certainly have their shortcomings, but many scholars agree that the region is characterized by a particular "collective psycho-cultural entity," and we can find some very general trends and commonalities among southerners. For instance, southerners (both black and white) are consistently more politically conservative and religiously oriented than the rest of the country. In Augusta, I found that nearly all Democrats, Republicans, whites, and blacks strongly supported prayer in schools. In addition, and perhaps most important, Augustans themselves (again, both black and white) often mentioned and appealed to their southernness—however elusive and complicated it might be.[14]

For African Americans across the country, the South figures prominently into people's identities. Not only do most have origins in the region, but it is also where the most brutal civil rights struggles took place. Literary theorist Houston Baker writes, "For the Black American majority of the 19th and early 20th centuries, 'the mind of the South' was critical to black personality, cultural, economic and political formation." Similarly, anthropologist Steven Gregory finds that for African American activists in Queens, New York, "southernness" described a bond among older activists and provided a shared foundation for their activism.[15] Being located in the southern region of this country has played both a practical and a figurative role in Augusta's history. Indeed, the city's political, economic, and social relations combined in both typical and exceptional ways to form the set of ingredients that constituted Hyde Park's particular blend of toxic stew.

From Cotton Marts to Golf Carts

Pre-Revolution to Post-Reconstruction

According to Augusta historian Edward Cashin, most of Augusta's first colonists were transplants from the Carolinas and Virginia, and the city held a "Virginia" attitude toward its slaves. In other words, in return for their servitude, slaves were treated benignly, and masters made sure they were clothed, fed, and housed.[16] But many whites also believed that blacks needed supervision, so local politicians issued laws controlling their behavior. For instance, local edicts prohibited blacks from having lights on after ten o'clock, smoking, walking with a cane, or using "saucy" language with a white person. In addition, blacks were not allowed at public meetings without permission or supervision. Cashin reports that in certain circumstances whites ignored such laws, and within the city's limits, slaves (who came from a few plantations on the city's outskirts) were able to hire themselves out for extra work. Some used their wages to buy property through white "guardians" and businesses. Such relatively lax conditions made Augusta home to a number of "free slaves"—those who were allowed to earn money, and those who had either bought or been given their freedom.[17] Between 1800 and 1851, African Americans established nine churches in Augusta.[18]

Yet the city was far from an oasis of black independence. For example, in 1819 a slave named Coco (or possibly Coot) was executed for leading a conspiracy to burn the city and to instigate a rebellion, indicating that slave conditions in Augusta must have been malignant enough to warrant revolt. In addition, after John Brown's raid at Harpers Ferry in 1859, a nervous city council debated whether to run free blacks out of town or to enslave them. It decided instead to crack down even further on their "liberties," preventing blacks from purchasing liquor or staying out after nine o'clock.[19]

At the same time, residential integration was common in the South, enabling blacks to live close to their white employers. Black and white Augustans also rode on the same streetcars (even if they had separate schools and churches, and had to wait behind whites at banks, post offices, and other places of business). During this time, black Augustans also established another thirty churches and five black newspapers. In 1882, the Methodist Episcopal Church founded Augusta's first institute of higher education for African Americans, Paine Institute, which later became Paine College. The Freedman's Bureau in Augusta was a large and active entity around this time, although, illustrating the limits of white "benevolence," whites ostracized the bureau, and the editors of the *Augusta Chronicle* criticized it for creating dissatisfaction among Negroes on the plantations. In 1865, three white vigilantes shot and killed an officer of the bureau. Moreover, a few years later, white hostility toward the Augusta Institute (a black seminary school) forced it to move to Atlanta, where it eventually became Morehouse College.[20]

As Reconstruction drew to a close in the late 1800s and federal troops left the South, race relations deteriorated further. The Republican Party abandoned southern blacks, and the Ku Klux Klan grew in strength nationwide, until most Georgia state office seekers "considered Klan membership a prerequisite to election."[21] The U.S. Supreme Court increasingly undermined the Civil Rights Act of 1875, culminating in the *Plessy v. Ferguson* decision of 1896, in which the Court sanctioned the "separate but equal" doctrine. Around this time, poor race relations in Augusta also influenced the state of race relations nationally. In 1891, the Richmond County Board of Education abruptly decided to close Ware High School, Augusta's first public black high school. Education brought rural blacks to the cities and prepared them for occupational opportunities that did not exist, ultimately leading to uncontrollable vagrancy, or so argued the board of education and other white leaders.[22] Augusta's black activists successfully rallied to oppose the school closing, but six years later the board once more announced that it was closing Ware. Again, black Augustans came together and filed a lawsuit that petitioned for the school's reopening. This time, both sides were tenacious. In 1899 the petition went all the way to the U.S. Supreme Court, where it became the precedent-setting case *Cumming et al. v. Board of Education of Richmond County*. Here the Court effectively refined its definition of equality in the separate but equal doctrine, insisting that no matter how disparate the outward conditions, a separate system was constitutional unless plaintiffs proved that it was inherently and deliberately unequal. In other words, the Court established a tacit understanding that separate did not have to mean equal.

Throughout the Jim Crow era, both local and state white leadership worked hard to maintain black disempowerment. Across the state of Georgia, whites bought Negro votes for a dollar each, plus barbecue and drinks before the election. To put an end to this practice, the state instituted a white-only Democratic primary in Georgia—a move that was well supported by white Augustans. Next, state lawmakers adopted an amendment to the constitution that specified that in order to vote, one must be a war veteran or a descendant of one, a person who could read and write a paragraph of the U.S. or Georgia constitution, or the owner of forty acres of land or property worth $500.[23] By the early 1900s, Georgia had effectively disenfranchised its African Americans. Here, both black and white Augustans followed the patterns of other southerners: whites worked to keep blacks politically subordinate, and blacks fought back in selected battles, often over education, highlighting how the history of black activism is as long as the history of oppression.

In economic terms, the turn of the century signaled tough times for Augusta. Once the second-largest inland cotton market, in the early 1900s the city desperately needed other sources of revenue. A statewide ten-year tax exemption for new textile and iron manufacturers facilitated the industrialization that was to pervade Augusta and other southern cities (and which would eventually lead to the "toxic stews" and "toxic donuts" scattered throughout the region). Augusta soon became a center for linoleum, rubber, paper filler, and the mining of kaolin. In addition, the city developed a secondary economy through scattered textile industries, an arsenal, and several hospitals that served rural areas. Shortly after the turn of the century, Augusta (now promoted as "the Lowell of the South") took its place as the largest cotton-milling center in the South.[24] It also developed a small tourist economy, advertising that its sandy soil and temperate weather made it an ideal winter resort destination. Here the city began to set itself apart by successfully diversifying its economy (although none of the new industries gave it the economic standing it had in the heyday of the cotton market).

While Augusta struggled to get on its economic feet, its black citizens struggled to establish and exercise their new freedoms and continued to build schools, newspapers, and churches. Yet the turn of the century's economic expansions had little effect on improving the lives of black Augustans. For instance, Richmond County tax records indicate that in 1910, fewer than 100 of Augusta's 18,344 black citizens owned property worth more than $2,000, and an overwhelming number lived in poverty. Not only were Augusta's blacks thus poor and disenfranchised, but also in the early part of the twentieth century, living among whites became impossible. For example, in 1913, local politicians passed a city ordinance that zoned residential districts according to race, cementing neighborhood segregation.[25]

In some ways, racist policies and practices strengthened black self-sufficiency and institutions.[26] Black churches, banks, insurance companies, doctors, and fraternal and social societies proliferated. Notably, after the Supreme Court debacle of the late 1800s, private schools became essential to African American education. As a result, black leaders (including several women) secured the assistance of white philanthropists and opened several private black high schools and institutes, as well as a life insurance company, a theater, and a community center.[27] Private enterprise also led to social segmentation within the black population, enabling a black elite to grow and solidify. Still, Augusta's black elite was only wealthy in comparison to the majority of other blacks, who skirted destitution. Throughout World War I (in which a number of black Augustans enlisted)[28] and the years immediately following it, Augusta remained a segregated city with a large, mostly poor black population.

Not Quite Slavery but Close: The Sharecropping System

As a meeting ground of rurality and urbanity, Augusta's agrarian context warrants some discussion, especially the sharecropping system, which developed in response to a rapidly deteriorating post-Reconstruction agricultural economy. For it was this exploitative system that most of Hyde Park's residents fled when they moved to the city. As the early part of the century wore on, the international cotton market fluctuated erratically, and overfertilization and the boll weevil destroyed much of the land.[29] Moreover, slavery's demise dismantled a credit system for white farmers that relied on slaves as collateral. This lack of credit, combined with Reconstruction costs and poor crop yields, left farm owners cash-poor. They could afford only to hire labor based on shared agreements, or tenancy farming, and a multitiered system of exchange took hold across the region. At the top of the system were land renters (those who paid owners for a piece of land and kept their entire crop); at the bottom were sharecroppers, who received a portion of the crop in exchange for working the land. That portion depended on the season's yield (as well as the landowner's honesty) and the kind of agreement the sharecropper had made with the landowner: sharecroppers sometimes had to give as much as three-quarters of their crop to landowners. Even so, landowners frequently overcharged tenants for supplies by advancing them money from the proceeds of their own crop and then charging them exorbitant interests rates on top of that. Tenants remained indebted to landowners for years and rarely saw any of the proceeds they were promised.[30]

Aside from the degree to which it exploited them, sharecropping was especially painful for African Americans because it represented unfulfilled postwar promises. Although in the early 1900s, nearly one-fifth of black farm operators owned land, by the middle of the Great Depression, most black farmers had ended up as sharecroppers—former slaves who had never received or been able to hold on to their "forty acres and a mule." As it turns out, much of that lost land was stolen. An Associated Press survey found that in known cases, 406 black landowners were cheated out of, and forcibly removed from, more than twenty-four thousand acres of farm- and timberland—all of which is now owned by whites or corporations. Land was taken by private citizens (who often used groups like the Ku Klux Klan to force families from their properties), as well as by state governments, which found pseudolegal reasons to revoke land claims and systematically denied or delayed loans to black farmers.[31] In the years following Reconstruction, land theft and a suffering economy left black farmers with little choice but to turn to sharecropping, which was not much better than slavery.

Landlords not only kept their tenants in debt peonage but also worked them as hard as they could and developed restrictive agreements, vagrancy laws, and other kinds of regulations that made moving from farm to farm illegal. In addition, owners constantly accused tenants of stealing, abandoning crops, or "impudence," which they punished with beating, whipping, lynching, or shooting. This violence was also used as a tactic for intimidating tenants and keeping them from running off to another farm. Yet historians have found ample evidence that many tenants did move, especially during abundant periods when labor was in short supply. Changing locations branded sharecroppers as unreliable and irresponsible, but it also gave them a modicum of autonomy, "fix[ing] the limits of white control."[32]

Between the Wars

Back in white Augusta, things improved during the flapper era. Textiles, cotton oil production, and clay dominated Augusta's manufacturing scene. City leaders capitalized on a nascent tourist economy, promoting the city as a sunny escape from the cold of the northern states. As part of this effort, Bobby Jones designed the course at the Augusta National Golf Club. Two new educational institutions—a private high school and a two-year college—were constructed. This growth and prosperity were sustained even during the Depression years, when Augusta managed not only to complete the golf course but also to acquire its first radio station, enlarge the university hospital, initiate airline and air mail service with New York and Miami, and build a lock and dam below the city.[33]

Such advances ought not to detract from the hardships that many Augustans, especially African Americans, faced during the Depression years. One of Augusta's most famous natives, soul singer James Brown, moved to the city in 1938 at the age of six to live with his Aunt Honey, who ran a house of prostitution in the heart of the "Terry" (for "Negro territory"). In his autobiography Brown states that back then, "Augusta was sin city; plenty of gambling, illegal liquor and a lot of houses like the one I grew up in."[34] As Brown notes, Georgia remained a dry state long after Prohibition was repealed. A long-standing system of payoffs sustained a corrupt local government and a host of illegal activities. However, even nonlegal businesses (like that of Brown's aunt) ran up against hard times during the Depression. Brown himself had to help support his family by shining shoes on a street corner. Life in the Terry and in black Augusta more generally was never easy, but during the 1930s it had become particularly bleak.

As America emerged from the Depression years, its black citizens lagged far behind, thanks in no small part to New Deal programs, which favored whites. Many of these programs were designed to help white farmers, who by the early 1930s earned 60 percent less for their products than they had before the stock market crash. New Deal agencies and programs paid farmers to reduce their crops. Although this strategy effectively raised commodity prices, it also meant that farmers hired fewer tenants. Such programs also subsidized mechanization, again shrinking the need for labor. In 1930, 80 percent of blacks living in the rural South were sharecroppers, tenant farmers, or day laborers; by the end of the decade, tenancy farming had declined by 25 percent.[35]

The good-bye to sharecropping was bittersweet, however, for the New Deal's worker-oriented programs had left southern blacks very much in the lurch. The National Recovery Administration (NRA), for instance, codified fair labor practices but exempted agricultural workers and domestics—the two categories that accounted for approximately three-fourths of southern black workers. In addition, the Agriculture Adjustment Administration (AAA), the NRA, and other programs gave local administrators considerable autonomy in applying New Deal programs and policies, which meant that subsidies and stipends were often channeled through white-dominated administrations and never made it into black hands. Although the NRA established equal pay scales for whites and blacks, agency officials looked the other way when southerners deliberately ignored these provisions.[36] For instance, Georgia governor Eugene Talmadge resisted equal wages for whites and blacks, giving Augusta bureaucrats a green light to distribute emergency relief jobs unequally. In 1940, blacks constituted 46.8 percent of the city's workforce but were granted only 31.7 percent of the local jobs made available by the federal government.[37] These conditions, combined with agricultural change, impelled four hundred thousand blacks to leave the South during the 1930s in a well-documented exodus.[38]

Although most of these migrants skipped southern cities and their racial caste systems and headed straight for the Northeast and Midwest, some ventured only as far as the closest metropolitan area. In Augusta, black farmers began pouring into the city's already crowded Negro districts, building makeshift houses on alleyways they named

"Thank God" or "Slopjar" Alley. James Brown confirms that in the Terry, "The streets were mostly unpaved clay and sand. Rows of cabins in alleys stood side-by-side with regular middle-class homes."[39] Indeed, residential segregation forced even middle-class blacks to erect homes in the midst of shanties.

Typically in the South, middle-class blacks were educators, and Augusta was no exception. Historian James Cobb reports, "of the 1.9% of blacks in Augusta classified as 'professionals' in the late 1940s, 91.6% were teachers."[40] The prevalence of black teachers played an important role in local black activism, which continued to center on educational issues. For example, in 1938, black Augustans finally won the battle they began in 1897 and reestablished Ware High School. A year later, however, the school board decided not to uphold its promise to make the new school a four-year institution. Over the next four years, Augusta's black activists filed numerous petitions and corralled enough support to revoke that decision.

By the time Augusta reached the middle of the twentieth century, it was thus a racially divided, socially segregated city on almost all counts. Its residential districts, schools, businesses, electoral processes, and social life were strictly structured in racial terms. Moreover, white elites consistently attacked and diminished black access to resources (such as higher education) that might enable blacks to achieve some power.

After World War II: The "Golden" Years

"The servicemen started pouring into Augusta in the fall of 1940," remembers James Brown.[41] Across the country, business boomed in the post–World War II year, and Augusta was no exception. However, the rising tide of industry was a murky one for Hyde Park residents. On the plus side, a bountiful economic climate nourished even neighborhoods like Hyde Park, and residents were able to establish stable work and community lives, which also fostered their early activism. On the minus side, Jim Crow was alive and well during this period, and the industrial nature of the growing economy carried serious environmental ramifications.

In 1941, the U.S. Army completed Camp Gordon, a fifty-six-thousand-acre training facility. Here again, Augusta's economic history was shaped by regional patterns. Between 1939 and 1941, military bases shot up on more than one million acres of southern land. Not only was land cheap, but also a southern climate was conducive to year-round military training. Although at the end of the war it looked like the influx of army personnel (and income) would decline, a few years later the army moved its signal training and military police schools to Augusta, sustaining Fort Gordon's role as a major local employer.[42] Municipal coffers continued growing into the next decade. New industries, including Continental Can, EZ Go Car Corporation, Wilson Shirt Company, and plants from GE and DuPont took their places in Augusta's greater metropolitan area. The robust economy brought jobs to both black and white workers. As usual, however, white workers were the primary beneficiaries of the new economy. In 1949, for example, the median black income in Augusta was $789, which was only 44.4 percent of median white income.[43]

At the same time, wartime experience *did* give some black men the opportunity to build savings, which, combined with an increase in jobs, enabled more blacks to own property. Arthur Smith described his father's experiences of returning to the Augusta area as a World War II veteran:

> Dad had served over six years in the military during World War II doing combat. Dad was a farm man before he left for the military and now he was a soldier, and he started working over at Clearwater Finishing Plant. . . . He bought this piece of land first. . . . That was the age when black Americans could finally buy a piece of land. I think there was a time in the early fifties, right after World War II. America was on an upbeat, an upswing.

As property ownership became more possible for many black families, the number of houses in Hyde Park (and in similar urban enclaves across the South) swelled. Of course, the location of these homes was proscribed by segregationist practices. Traditionally, southern cities set aside undesirable land on the edge of the city center (and near swamps, creeks, railroad lines, and industry) for black housing sites.[44] Hyde Park was one such neighborhood, and a disproportionate number of Augusta's burgeoning industries surrounded it, already simmering what would become a toxic stew.

By the end of the 1940s, Babcock and Wilcox (later, Thermal Ceramics) was the fifth-largest employer in the city of Augusta.[45] Merry Brothers Brickyard, on the neighborhood's opposite edge, was also booming

CE behind his bar, 1999. Photo by Maryl Levine.

and hired a good number of Hyde Park men. In the mid-1950s, approximately one hundred homes in Hyde Park (all of which were owned by blacks) filled three streets, and employment was at nearly 100 percent. In keeping with Augusta modes of black self-sufficiency, residents built three or four churches on as many streets and opened several businesses. For example, Gordon's on Golden Rod Street was both butcher shop and bar. Hammer's and CE's sold beer and groceries, and a man named Nelson opened a restaurant. Most of these businesses operated at night and on the weekends, after their proprietors had already finished long days of factory work.

At the time, Hyde Park was contiguous with Aragon Park, the smaller neighborhood next to it. Many residents of the two neighborhoods were kin and shared resources, such as child care and food from backyard gardens. Aragon Park residents also relied on Hyde Park's churches, convenience store, barbershop, and other small businesses. Work was hard and money tight, but community was even tighter. Charles Utley remembered,

> We had no cars. Very few cars in this area. You had a car, you cherished it. We had a community television. There were maybe two televisions on Walnut Street for the whole street. And we would always watch tele-

> vision on Sundays. And the primary show would be the *Ed Sullivan Show.* So we would all get together.

Mary Utley (Charles's mother) played a major role in maintaining the unity her son describes. She organized transportation to take seniors to doctor appointments, formed programs for local children, and planned various neighborhood events.

Robert Striggles lived in Hyde Park from the age of six to forty-six. He recalled,

> So it really wasn't a bad area to live in up until about twenty years ago. . . . Before that everyone in this area owned their own home. . . . Summertime, the activities that we had out here, it was nothing but a ball field. Where the recreation center is there? Right in that area. . . . Basically you had baseball, you had horseshoes and that's about it. But this way the neighborhood came. And not only the kids was there, when we had games, the parents was there. And that really kept us together.[46]

At the same time, Hyde Park residents struggled to make ends meet and to deal with the daily injustices of the Jim Crow era (and, surely, not everyone got along *all* the time). What is important, however, is that in residents' views, the 1950s and 1960s cemented community cohesion, which remained intact in the decades that followed.

In 1954, however, the city ran a new highway right between Hyde Park and Aragon Park. Curving around three-quarters of Hyde Park, the highway further sealed it off, making it impossible to leave the Park without crossing at least four lanes of traffic. Thus, traveling on foot between Hyde Park and Aragon Park now meant risking physical danger. Residents asked municipal officials for a footbridge or at least a traffic light on Gordon Highway. After approximately ten years and several pedestrian accidents, the county installed guardrails along the portion of the highway running between Hyde Park and Aragon Park. Yet the guardrails only further materialized the separation between the neighborhoods. Robert Striggles remembered,

> It really cut us off, the guardrails did. And it was dangerous anyway, walking back and forth across there, in the evenings making a turn to go up the highway. So it really split the community. Matter of fact, we complained at first.

This highway situation is not uncommon. Across the country, the 1950s marked a period of "urban renewal," in which old structures were bulldozed to make way for new, modern buildings and other amenities. Often this process decimated older black neighborhoods, either forcing relocation or creating boundaries that separated blacks from the rest of the city.[47] What is perhaps less common here is that Hyde Park residents did not take the disruption of their community passively.

Highway imposition was not the only thing that was contested in Hyde Park. Indeed, close-knit community life mitigated, but certainly did not negate, the effects of poverty and racism on Hyde Park residents. Until 1970, the neighborhood's streets were unpaved, city water and sewage lines did not exist in the Park, and there were no streetlights. Floods were so bad that residents occasionally had to leave the neighborhood by boat. Families with canoes would paddle from house to house, taking parents to work and kids to school. (Of course, for those who worked or went to school within neighborhood boundaries, floods meant a day off.) In addition, although employment was high, it was also dangerous. Job ceilings for southern African Americans limited their opportunities, and most were relegated to menial and risky factory jobs. For example, David Jackson's father smashed his leg while working at Merry Brothers and was out of work for several years. Earl Palmer suffered a back injury at the Federal Paper mill while still in his mid-thirties and spent the rest of his life on disability.

In the 1950s, some unionizers came to the Park to organize its workers. One of Hyde Park's first residents, Lola Kennedy, recalled,

> [The unionizers] come in and decided to help the working folk. Cause people wasn't paying them nothing. They paid them something, but it wasn't much, like two dollars. Things like that. Some of [the industries] did start giving [their workers] a little more afterwards, but it took them a little while to do it. Then you had to pay [the workers] so much an hour.

Hyde Park residents thus established a pattern of affiliating with other professional groups that continued into its later environmental struggles.

In the 1950s, Hyde Park founded itself as a tight-knit and civic-minded neighborhood. At the same time, residents were well aware that they received unequal pay, job opportunities, and city services (among other things). When those inequalities threatened their immediate community life, or their health and safety, they rallied together and voiced their opposition. Through these early struggles, Hyde Park residents envisioned the new possibilities for which they would wage long, protracted battles over the years to come.

Struggling to Say Good-Bye to Jim Crow

Politically, the mid-1950s bore enormous changes for the country as a whole. The cold war was under way, and McCarthyism was in full swing.[48] By the mid-1960s, violent racial clashes would rock southern cities like Atlanta, Mobile, and Birmingham. Augusta, however, remained relatively quiet. Even Hyde Park residents, who had demonstrated their activism in the past, did not join citywide or national civil rights protests. With a few exceptions most residents reported that they were too busy trying to earn a living. One woman said, "I had to work and that's why I couldn't do marches." Another told me, "I worked from twelve years old up to seventy-two. I didn't have time to protest. I worked every day."

Even though Augusta was not a hotbed of overt activism, a few concerted civil rights actions did take place there, illustrating several pertinent points. First, even sporadic and scant activity demonstrates that economic, political, and social relations in Augusta were polluted by racial inequality, and black Augustans were not content with the status quo. Second, activism in Hyde Park centered on neighborhood issues such as infrastructure and city services, highlighting how battles for city services were as essential as marches, sit-ins and demonstrations in boosting solidarity and dignity, and in working collectively toward change.[49]

According to news reports and local historians, the city of Augusta experienced six demonstration-like events during the 1950s, 1960s, and 1970s. The first occurred in 1957, two years after Rosa Parks famously refused to give up her bus seat, when Bessie Abrams chose to sit in the front of the bus after a long day at work. Abrams was promptly arrested, and when her bail was not posted in a reasonable time, a committee formed to raise the money. In May 1960,[50] a group of black students tried again to board a bus and sit in front of white riders.[51] Seven months later, 60 students sat down at white lunch counters in four downtown stores. In one store, 16 students were arrested, and minor violence ensued. The following year, two Paine College students were prevented from trying to enter a white church. In 1962, 150 Paine College students sat at an all-white lunch counter, this time sparking neither violence nor arrests. That same year, a local store's refusal to hire black workers led black activists to picket for several days. Picketing devolved into rock throwing, and in a black neighborhood someone shot a gun at a white-driven car.[52]

Unsurprisingly, white Augustans viewed race protests unfavorably, especially early on. In 1960, for instance, the *Augusta Chronicle* referred to local demonstrations as a "disgrace." Two years later, however, the *Chronicle* (usually a conservative voice) publicly supported school integration—a decidedly unpopular opinion in the Augusta area. In fact, the desegregation of Augusta's schools proved to be the stickiest aspect of integration for many white Augustans, as well as many Georgians.[53] Although the U.S. Supreme Court rendered school segregation illegal in 1954, the Georgia State Board of Education announced that it would integrate schools only if forced to do so. It then declared that any teacher caught teaching racially mixed classes would be suspended for life and demanded that black teachers withdraw their memberships from the NAACP.[54] Georgia governor Marvin Griffin and Georgia state senator Roy Harris from Augusta traveled to Little Rock to convince Arkansas governor Orval Faubus to defy the court's order to integrate Central High School in 1957.[55] Two years later, Ernest Vandiver replaced Griffin as governor, basing his campaign on the slogan "No not one!"—meaning that no black child would attend a white school.[56]

Local white integrationists remember the zeal with which their opponents strove to maintain segregation. Ernestine Thompson, a professor at Augusta State University, recalled,

> In Richmond County, it was really volatile for a couple years. The school board was violently opposed to integration. . . . When they started bussing, when they started integrating, the chairman of the school board told parents to keep their children home.

Thompson herself became embroiled in controversy after writing a letter to the editor in the *Chronicle* opposing the activities of a local anti-integration group. She recounted the angry reactions of some Augustans:

> Every day in the afternoon, I'd get a call. We got hate mail until the day we left [the neighborhood]. Some of it came through the mail and some of it was put in my box by neighbors. The chairman of the school board was in that neighborhood and called all the people in our carpool and said, "I don't think you want your children to ride with her."

Thompson's statement emphasizes not only the vehemence with which anti-integrationists stood their ground but also the degree to which in a "big, small town" like Augusta, such political wars become highly personal.

In 1962 four black students applied to Augusta College, but all were turned down for failing to complete necessary testing before registering.[57] In 1964, in response to a lawsuit, the Richmond County Board of Education agreed to integrate grades one to three, yet only ten black students enrolled in four schools that year. Integration proceeded at a similarly slow pace until 1970, when a federal judge ruled that the Richmond County Board of Education had deliberately failed to integrate its schools and called on two outside mediators from the Northeast to draw up an integration plan.[58] Finally, in February 1972, Richmond County implemented its first real phase of elementary desegregation. Seven months later, it began on secondary schools. As Thompson observed, "Augusta's never on the forefront of anything. It takes a while to get things going around here." Although some large southern cities that witnessed economic growth during the 1950s and 1960s integrated comparatively rapidly (hoping to appeal to northern industries),[59] in Augusta, cultural and political ideas sometimes took priority over economic concerns.

Augusta's racial tensions did not actually come to a head until May 11, 1970, when the city witnessed its first and only race riot. For local black activist Reverend Robert Oliver, the outbreak of violence had been a long time coming. Groundwork for the riot, he explained, was laid in December 1969. Oliver and some fellow Korean War veterans were sitting in a bar on Ninth Street (now known as James Brown Boulevard) in what had once been the "Terry." Suddenly, a young man ran into the bar and announced that Grady Abrams, one of the city's only two black councilmen, had been arrested. Several days earlier, Abrams's nephew had tried to pass off one of Abrams's checks at a local grocery store, and the check bounced. On hearing that, Abrams went to the store to make the check good, but the storeowner had the police promptly arrest him. Oliver recounted,

> [Grady Abrams] was in the back of the squad car and pulled out a book and started reading. The book was *The Life of Malcolm X*. This is 1969 Augusta, Georgia. They almost had a heart attack. Well, we were sitting in the Club DeSoto and a little fellow came and said, "Man, they just arrested Grady." We said, "Let's go get him." Out of two hundred people in the joint, only ten of us went. That's how we came to be called the Committee of Ten. All of us were veterans.

The fact that the "Committee of Ten" consisted entirely of veterans is not surprising. Experiencing integration during the Korean War galvanized many black veterans into racial activism.[60]

According to newspaper reports, Abrams claimed that police officers referred to him as "boy" several times, and when he took out the book, one of them said, "That's what's wrong with you—you've been reading that Malcolm X—and watching too much television."[61] Despite their efforts, Oliver and the Committee of Ten failed to get Abrams out of jail. They did, however, alert the local leaders of the Southern Christian Leadership Conference, the NAACP, and other black leaders to the situation. Pressure from these groups prompted a probe into Abrams's arrest. Although the Abrams incident did not directly incite public demonstrations, it exemplifies the treatment that black citizens received from the police and other municipal employees in the late 1960s. It also illustrates some of the reasons that racial unrest was serious enough in 1970 to lead to a May riot.

On May 9, Charles Oatman, a sixteen-year-old African American, who some claim was mentally retarded, was murdered in the newly integrated county jail where he was being incarcerated. Word soon got around that, although official reports said the death was accidental, a host of gruesome bruises, including fork marks and cigarette burns, had been seen on the boy's body. The news ignited black Augustans' frustrations over their unfair treatment by the police and brutal conditions at the county jail, where a disproportionate number of blacks were (and continue to be) incarcerated. Reverend Oliver explained,

> [The riot] was on May 11, 1970. It was more like a gang fight. Brian Oakman [*sic*] had been arrested at fifteen years old for burglary—for

robbing a gas station down in East Augusta. They arrested Brian Oakman and put him in jail with grown men. We were shown Brian's body. It was down at the Mays funeral home. They said he fell out of bed, but he had cigarette burns all over his body. He had a welt in his head right there. They said he fell out of bed: He was raped and tortured.
MC: So police brutality was pretty bad at that time?
RO: Oh yes. Oh yes. They're gonna deny it, but they're damned liars. Make sure you get that—they're damned liars. See that knot between my eyes up there? That's police.

It is important here to note Oliver's insistence that "they" would deny that police brutality existed. Other black activists often agreed that whites in Augusta denied racial tensions, let alone abuses. As Ernestine Thompson pointed out, right up until the day of the riot, editorials in the *Chronicle* consistently celebrated the city's harmonious relations.

But reactions to the teenager's death indicated that racial tensions had snapped. Local black activists initially organized a protest at the municipal building. According to the *Chronicle,* after one man tore down and burned the Georgia state flag, about three hundred people marched toward Broad Street and began overturning vending machines and trash cans. As dark approached, the crowd grew, and some people began to loot and burn local retail stores. Six black men were shot and killed in the melee, sixty-two people were injured, and rioters incurred more than $1 million in property damages.[62] Georgia governor Lester Maddox ordered two thousand National Guardsmen to control the city, and troops remained in Augusta for approximately the next seven days.

Although the riot wreaked havoc on both black and white Augustans, it did have the positive effect of inspiring Augusta's black leaders to take greater initiative in working for change. Singer James Brown flew into town and went on local radio to plead for an end to the violence. In his radio address, the outspoken Brown remarked that many existing black leaders "could not be called black" because they "don't represent the black community."[63] For Brown, the pastors of Augusta's main African American churches, its school principals, and its smattering of local black attorneys had become too accommodating to a white power structure. In a different take on the passivity of black leaders (and in keeping with their insistence on the absence of racial tension), Augusta's white leaders blamed the riot on outsiders. A popular rumor, which, although incorrect, continued into the twenty-first century, reported the Black Panthers had called for the riot, and an airplane full of them had landed at a local airfield, ready to march on the city.[64]

In any event, the riot provided a wake-up call to both black and white leaders. Almost immediately, the mayor instituted a plan to appoint more blacks to city council committees, to "very promptly" implement fair employment guidelines for city employees, to request that the chamber of commerce encourage investment in black-owned businesses, and to create a biracial human relations commission.[65] While the plan quelled tensions in an immediate sense, it also represents both the promise and the failure of Augusta's power structure to address continued racial tension in any serious way.

The New South: Economics, Politics, and Promises

The year 1970 was a landmark date in Hyde Park and signaled that change was afoot throughout the South. After two years of struggle and agitation, Hyde Park residents finally won their battle for water, sewer, and gas lines, as well as paved roads, streetlights, and a community center. But at the same time, the neighborhood was becoming increasingly boxed in by industry. The Dixieland Junkyard, a Georgia Power plant, Thermal Ceramics (formerly Babcock and Wilcox), a car junkyard, a brickyard, and a wood treatment plant now surrounded it. By 1970, the Gordon Highway had become a major thoroughfare shuttling traffic between rural areas south of Augusta and the growing city. The expansion of the highway and even the implementation of water, gas, and sewer lines represented a new economic upswing.

Thanks to the "New South," an unofficial political and economic campaign begun in the 1940s,[66] the region also began pulling out of an economic downturn that had lingered since Reconstruction. In the 1930s, the federal government nationalized wage and labor standards. Until then, the southern economy operated under its own rules, with self-contained and self-regulated businesses that depended largely on agriculture. New nationalized standards coupled with persistently poor economic conditions inspired southern leaders to end their isolationist tendencies and seek northern investment in the region. Between 1950 and 1978, corporate taxes fell from 85 percent above to 13 percent below those in the rest of the country.[67] These lower taxes signaled a concerted effort to attract northern business. State governments also enticed new business by establishing tax exemptions for manufacturing plants and low-interest loans. Importantly, these state subsidies also included relaxed enforcement of pollution standards and environmental regulations. Moreover, the South offered a nonunion climate and plenty of low-wage workers. Efforts to attract business met with substantial success. Between 1970 and 1980, population numbers grew at faster rates in the South than in the nation as a whole, and between 1960 and 1985, seventeen million new jobs found their way into the southern economy. Per capita income also grew at rates far above the national average.[68]

However, these numbers tell only one side of the story. As many historians point out, rather than spreading economic development, the New South actually meant replacing one economy with another. In other words, while high-tech industries grew, and per capita income increased, in many places low-wage and manufacturing jobs dried up. Thus, the South's population influx consisted mainly of highly skilled, well-educated migrants who raised its per capita numbers. Unskilled workers, on the other hand, left the region in search of better jobs, and many of those who remained fell into poverty. Once again, this increasingly two-tiered economy affected blacks far more than whites. Between 1950 and 1970, for instance, black teen male employment in the South decreased by 27.4 percent, as compared with white teen male employment, which declined only by 4.8 percent.[69] As industries globalized, rural areas especially suffered from a serious loss of jobs in textile industries. Most of these areas are part of what is known as the "Black Belt," or nonurban parts of Alabama, Mississippi, Louisiana and Georgia that have large black populations.

At the same time, the black exodus that had occurred throughout the post–World War II period reversed itself in the 1970s and 1980s, and the South's black population grew by 23.4 percent.[70] These blacks from the North were better educated and had higher occupational status than native black southerners. If blacks reaped any benefits from the New South economy, it was the northern black immigrants, who located mainly in the largest cities. For instance, while throughout the 1970s and 1980s, the number of manufacturing jobs in the Augusta area stayed relatively constant, hovering between 34,000 and 35,000, in Atlanta between 1976 and 1980, manufacturing jobs shot from 125,000 to 146,000. Augusta's per capita income also reflects this unevenness—in 1950 it was only $300 below Atlanta's, but by 1972 it was more than $1,000 short of Atlanta's. Accordingly, between 1970 and 1990, Augusta's black population grew 12 percent, only half as much as the South's overall.[71]

In Augusta, migration patterns also diverged from regional trends. A number of Hyde Park residents, for example, moved north for some period of time, but unlike other southern African Americans, few stayed very long. David Jackson (and one of his brothers), Louvenia Calloway (and one of her brothers), Johnnie Mae Brown (and her stepsister), Sam Jones, and Melvin Stewart all moved to various parts of the Northeast but stayed only a few years before returning to Augusta. Their reasons for returning were similar: they found that they made more money up North, but that the higher cost of living prevented them from either building savings or sending money back home. Life up north, they agreed, was "too expensive" and "too fast." Often, these people found they could be of more help to their families by being physically present than by sending what little extra money they scraped together.[72]

In the meantime, thanks to the advent of a few new companies and the expansion of others (including Procter and Gamble and RUTGERS Organics Corp.), and to the growth of Augusta's hospital industry, the city's economy grew at a slow but steady rate throughout the 1980s and 1990s. Between 1992 and 1997, per capita income in the county rose 23.1 percent—only 3 percent less than for Georgia as a whole. Still, per capita statistics do not give a full picture of income in the area. For instance, in Richmond County, 19.6 percent of people lived below the poverty level in 1999, compared with 13 percent in Georgia overall.[73] Of those impoverished people, an overwhelming majority were (unsurprisingly) black.

What the Past Produces

In the 1980s and 1990s, differences between rich and poor—and between white and black—were as pronounced as they ever were. In 1989, 30.8 percent of blacks in Richmond County earned less than $10,000, compared to 13.6 percent of whites, and only 9.3 percent of blacks earned more than $50,000, compared with 21.4 percent of whites.[74] After moving operations to the South, manufacturers now moved them overseas. Combined with technological change, this trend catalyzed massive layoffs of both white and black workers. However, white workers, who did not face the racial discrimination that black workers faced, fared better in finding alternative jobs.[75] Neighborhoods like Hyde Park, whose population relies on low-skill labor, are especially hard-hit in the process of deindustrialization. For instance, in the 1950s and 1960s, Babcock and Wilcox industrial ceramics plant was "the place to work" in Hyde Park—it was located right in the neighborhood, pay was decent, and jobs were plentiful. At its peak of production in the late 1960s, the plant employed 1,500 people. But by 2004, that number decreased to approximately 450.[76] In this case, downsizing resulted from major changes in steel production, which no longer relied on ceramic bricks. By 1999, such reductions in manufacturing jobs were apparent in Hyde Park, where the unemployment rate was 18 percent (unemployment rates in the county averaged approximately 7 percent between 1996 and 2000) and the poverty rate was 67.4 percent[77] (compared with approximately 19.6 percent in the county).

Political issues also continued to follow a color scheme. For example, in 1998 former newscaster Bob Young, a white Republican, ran for mayor against Ed McIntyre, a black Democrat. In 1981, McIntyre had been elected Augusta's first black mayor; after two years in office, however, FBI agents arrested him on bribery and extortion charges, and he served fourteen months in federal prison. Blacks in Augusta generally forgave McIntyre, and many told me they believed he was innocent and had been set up. But throughout the 1998 mayoral race, numerous polls and interviews quoted both black and white voters saying, for example, "The color of [the candidates] doesn't have anything to do with [the election] with me."[78] When election day came, however, Young won 55 percent of the votes and McIntyre won 45 percent. Not coincidentally, 55 percent of Augusta's voters are white and 45 percent are black; Young won in all majority white wards and McIntyre won in all majority black wards. In the two districts that are racially balanced, Young won one and McIntyre won the other. Despite Augustans' insistence to the contrary, the mayoral race was clearly about black-white issues.

The racial politics that have guided Augusta's economic development have, in turn, directly hurt its poor minority neighborhoods. For instance, over the years, Augusta's success in attracting business means that it now houses approximately 35 chemical-producing facilities.[79] These industries tend to be clustered near both low- and middle-income black subdivisions. The lackadaisical environmental regulations used by southern states (particularly Georgia) to lure industry have thus

Goldberg Brothers scrap metal yard, 1999. Photo by Maryl Levine.

positioned Augusta's black citizens on the front lines of toxic hazards.[80] In Hyde Park, such unchecked growth in industry also allowed that the scrap yard bordering the neighborhood to expand until spare tires formed pyramids that rose higher than the roofs of neighboring houses. By the 1990s, so much debris overflowed the yard that it threatened to block one of the neighborhood's only access roads.

The junkyard represents only one aspect of the trouble associated with living in a "toxic stew." The ceramics factory often left a white powdery dust on the cars of nearby residents, and its smokestack penetrated any view of the sky. The crisscrossing metal towers of the Georgia Power plant station formed an unwelcome industrial sculpture on the neighborhood's southern edge. Odiferous water filled the ditches in heavy rains, and car tires bumped over train tracks each time they entered or left the neighborhood. The chemical companies and plants around Hyde Park not only had contaminated its air, water, and soil but also had physically isolated the neighborhood from the rest of Augusta. Thus, in 1998, the uneven development of the New South had left Hyde Park in fairly dire straits economically, physically, and ecologically. The neighborhood suffered from job loss, the neglect of local politicians, and the ill effects of chemical-producing industries.[81]

Yet Hyde Park was also full of the complexities and contradictions that characterize any community. On a physical level, its landscape amalgamated metropolitan and rural characteristics. Grass grew high (if, in some areas, untamed), and tall pecan trees spread their branches across large yards. Most houses were freestanding and wood framed, interspersed with the occasional mobile home, or "trailer." Nearly every home had a wide and well-used country-style porch. Some were brightly whitewashed, trimmed in red or blue, and bordered by flower-filled lattice, while others had peeling paint, dirt yards, and porches cluttered with odd pieces of furniture.

Of the roughly two hundred houses in Hyde Park, approximately 10 percent stood vacant.[82] Longtime resident Betty Hall explained,

> Now there's a lot of empty houses. People have moved out, some of 'em have died out. People don't move, them that could get out, they get out on account of the place was contaminated. Them that was working and could get out, moved to some other place. But you see when you old as us, you aren't working and you can't move out.

Most unoccupied homes had been left to crumble, and crack addicts and homeless persons often squatted in them.[83] Other houses that appeared uninhabitable were actually filled with extended families.

While the majority of families struggled to keep up their homes, some families had typically middle-class occupations and lifestyles and chose to stay in the neighborhood despite its problems. Several people, for example, had aging family members in the neighborhood and preferred to live close to them. Others refused to sell their homes at significantly reduced property values and decided to stick it out until they were relocated or the contamination was remedied. Some residents whose families were relatively healthy refused to admit that the contamination was real.[84]

Another important reason for families choosing not to leave the Park was the simple fact that they had grown up there, and their neighbors were kin—both literally and figuratively. In 1998, a neighborhood survey reported that two-thirds of residents surveyed had lived in Hyde Park for more than twenty years, and one-third had lived there for more than thirty years.[85] Many people were known by affectionate nicknames such as "Mule," "Bull," "Snake," "Blossom," "Juicy," "Man-Man," or "Lil' Mama." Neighbors often acted as godparents for each other's children, and adopted one another into their kin groups. David Jackson described his relationship with one of his neighbors:

> Well, right across the street, a lady named Vi . . . my kids call her Aunt Vi. She's not their original aunt or anything, but anything that they need from babysitting—they can go to her when she can't come to us sometimes. She'll be doing something so she'll take them home. We try to live like a family.

Just as they had since the 1950s, residents continued to share their resources in the 1990s by exchanging child care, rides, and other favors.[86] Certainly, relations between neighbors were not always rosy. Over the years some people had disputes and nurtured grudges as a result. And, as is true anywhere, some people simply disliked one another.

One thing nearly all residents had in common was religion; most were Baptists, although a few practiced Pentecostalism or Catholicism. No matter what their denomination, nearly all residents attended Bible studies on Wednesday evenings and church on Sundays. Some went to Christian Fountain, the only one of Hyde Park's three churches that was still active in 1998.[87] More often, though, residents went to churches in other parts of town and returned to their "home" or "membership" church (usually in Waynesboro or Millen) at least one Sunday out of every month. Betty Hall explained,

> My membership church ain't [Christian Fountain]. But any Sunday, I'm not at my membership church, I'm there. I go there more than I go to my membership church because we have church once a month or every fifth Sunday at my membership church.

Skipping a monthly service inspired much guilt, and church anniversaries and other special events were rarely missed. Attending services at a "home" church was a crucial part of staying connected to the past and straddling urban and rural worlds.

The character of life in Hyde Park bucked many common stereotypes about living in a "ghetto." Politicians often cavalierly blame the problems besetting neighborhoods like Hyde Park on a lack of community life. Just after his inauguration in January 1999, for example, Augusta's mayor, Bob Young, launched a campaign to revitalize inner-city neighborhoods. According to Young, the way to solve crime in these (and all) areas was to

> get involved in the lives of our neighborhoods, get to know our neighbors, go to jury duty. Make our neighborhoods neighborly. Build a sense of community. Then the crime elements will be run off.

For Young, "un-neighborly" residents create their own problems. But Hyde Park residents, unlike residents of many typical suburbs, shared resources, had known their neighbors for almost their entire lives, and remained deeply connected to their rural and religious roots. Moreover, and perhaps most important, Hyde Park residents had a long history of coming together and fighting to improve their lives.

The social and ecological pollution that Hyde Park found itself mired in by the end of the 1990s did not result from a lack of community cohesion or concern. Rather, it stemmed from the fact that historically, its residents have been last hired and first fired; that they often have had to quit school to help support their families; that their schooling has historically been substandard; that they are surrounded by toxic industries; and that they have had to fight long and hard for the kinds of resources that middle-class white neighborhoods take for granted. If anything, Hyde Park's problems have made it a closer community, one that certainly does not take its neighborliness for granted.

Conclusion

To understand black activism in Augusta, Georgia, it is not enough to have some knowledge of "the South." Rather, we must comprehend the various ways in which Augusta's history has both followed and diverged from southern history. Like most cities, Augusta is a bundle of contradictions. It is a sprawling, populous city with a small-town feel. It is Georgia's second-largest metropolitan area, but not one of its most well known. Even Augusta's main claim to fame has contradictions. Tiger Woods was the first black man to claim a major golf victory there back in 1997, but more recently Augusta National has come under fire for not allowing women to compete in the tournament. As people across the country scratch their heads and wonder why Augusta

National's chairman, "Hootie" Johnson, refuses to change club rules (despite his reputation for being fairly progressive), we get inklings of the individuality and obstinacy that characterize city inhabitants and the city itself.

In economic terms, Augusta has remained stubbornly neutral, experiencing neither the highs nor the lows of regional trends. After the devastating effects of the boll weevil and Reconstruction on the cotton economy, the city diversified its economic base, but it never fully bounced back. As a result, Augusta did not experience a major bust during national economic downturns; but, by not finding a product to replace cotton, it never really boomed either. During the post–World War II years and the heyday of the 1970s New South, the city did not achieve the economic prominence of southern centers such as Atlanta, Charlotte, Richmond, or Raleigh-Durham, although it did manage to maintain a slow, steady rate of growth. Today the city's economy is neither robust nor declining. When we look at who has benefited from the city's economic growth, however, we find more extreme statistics.

Whites in Augusta have overwhelmingly held on to economic power, as well as social and political control. Here again, we find a city full of contradictions. For instance, "official" versions of city history work hard to establish a veneer of racial tolerance, insisting that Augusta has had a more equitable racial trajectory than its southern sister cities. Yet black activism throughout the city's history belies that idea. Although they might not have led the South in lunch counter sit-ins, bus strikes, or voter registration efforts, as sure as racism has remained alive and well in Augusta, Georgia, black activists have been there to fight it. Finally, although the city developed a set of race relations specific to its locale, Augusta *does* share some of its history with other southern states, and that shared history frames a certain local ethos among blacks and whites. In the end, it is both their unique and their typical experiences that interweave into the multilayered identities of the people in this ethnography, who see themselves as southerners, as Augustans, as Hyde Park residents, as Christians, as African Americans, and as activists.

Four

Strange Fruit

Stepping onto Louvenia Calloway's porch, I said hello to two of her neighbors, who crouched there, skinning fish they had just caught in the Savannah River. A tall, thin woman of sixty who looks at least fifteen years younger than her chronological age, Louvenia greeted me with a smile. She ushered me into her front room, and we each sat down on one of two spotless white sofas arranged in an L shape around a glass coffee table with a gold base. Over the coffee table, a gold chandelier hung from the ceiling, and beneath it lay a white carpet.

The living room's formality and pristine condition belied the fact that it served three adults, one five-year-old girl, two birds, two cats, a dog, and fish. Across the hall, things looked a little less surprising. The dining room turned playroom was strewn with toys. A small girl knelt in front of a TV, trying very hard to be quiet while we spoke. Louvenia settled back on the sofa as I set up my tape recorder. This was our second interview, and although Louvenia was usually shy and soft-spoken, she seemed far more comfortable than the first time around. I began by asking for an update on her children's health.

In the early 1990s, Louvenia's oldest daughter, who attended Clara E. Jenkins Elementary School, fainted a few times at recess and shortly thereafter developed enlarged lymph nodes on her neck. A biopsy of the nodes revealed that her condition was benign, but that was only the beginning of Louvenia's troubles. Soon her second child, Terrence, also got sick at Jenkins Elementary. Louvenia recalled,

> *At first I thought it was just an infection or virus or something he had because he showed no kind of signs of sickness, no more than running a temperature. But at night it would go real high—to 105 or 106. Then I had to call the doctor and she sent him to a specialist, but they couldn't seem to find out what was wrong with him. He went from, he was*

> *weighing about 105, 110, somewhere along in there, to 49 pounds. And he was so tiny until he couldn't even put on a pair of shoes to walk in because they were too heavy for him.*

With her young son in and out of the hospital, Louvenia quit her nursing job and devoted herself to her children full-time. Because she did not yet qualify for retirement, she signed up for public assistance to pay her bills and obtain medical insurance.

Eventually, enlarged lymph nodes appeared on Terrence's neck. After a biopsy, he was diagnosed as having T-cell lymphoma, which is extremely rare, especially in black children. The doctor told Louvenia the cancer was caused by toxic chemicals. Louvenia remembered being unable to prevent her children from eating pecans off the backyard tree. She connected those pecans to their illnesses (although she never sought legal action beyond the HAPIC lawsuit). After a year and half of chemotherapy, Terrence's cancer went into remission.

Unfortunately, health problems were still not over for the Calloways. In 1997, Louvenia had a baby daughter, Tanisa. As an infant, she began suffering from severe ear infections and asthma. At the age of two Tanisa had not yet uttered a word. Louvenia found out that a specialist from Atlanta was coming to Jenkins Elementary to conduct vision and hearing tests on neighborhood children. The specialist told her that her child could not speak because she could not hear. Louvenia recalled, "They put tubes in her and we went outside and there was a loud sound and she jumped so we knew she could hear." The tubes kept falling out, though, so the doctors decided to remove Tanisa's adenoids. Finally, she began to speak.

Just when things were calming down for her children, Louvenia herself became very depressed. "I just wouldn't let go and when it hit me, it hit me real hard," she said. After some treatment, she came out of her depression. Both her older children were in college in Augusta by then, and in 2003, Louvenia reported happily that Terrence was "fat" and engaged to be married. But Tanisa "still ha[d] a long way to go." Her language skills lagged behind those of other children her age, and she still suffered from severe allergies. Louvenia often had to bring her home from school because of asthma attacks. Some of her doctors suspected she was autistic. Tanisa's fragile health (not to mention the absence of a family car) prevented Louvenia from returning to work. But, *on the bright side, she told me as she walked me onto the porch and lit a cigarette, in only two more years she would finally be able to collect Social Security.*

Early on a Saturday morning in 2003, I stood on David Jackson's slightly slanted porch and knocked on his door. I had not seen David in over a year, and I looked forward to catching up with him. A medium-sized man in his late fifties with closely cropped hair, David was known for his beautiful singing voice. David greeted me at the door wearing his usual outfit—black pants and a black button-down shirt. After waving hello to his neighbor, "Wheaty" Brown, he showed me in and motioned for me to sit on a long sofa while he took a chair against the wall. In between us a big, round coffee table was covered with framed photographs of the Jacksons' children and loved ones.

David had spent most of the last fifty-three years in this house on Walnut Street. For a good part of that time, he had to deal with the fact that, with each passing year, Goldberg Brothers scrap metal yard crept closer to his property. When it rained, pools of oil and sludge collected in David's backyard, and for years, he imported fresh soil to cover the smelly, strange-looking dirt he found there. In the early 1980s, the scrap yard's owner demolished Jackson's backyard shed but refused to accept responsibility. Jackson took the case to small-claims court. "I didn't have a lawyer, but I had everything together. The judge gave me credit for having everything so well organized, but he decided it was an 'act of nature.' Wasn't no act of nature; [Goldberg] tore my building down," David insisted.

For David, the tearing down of his shed was about as natural as the reasons his father's once-plentiful garden produce stopped growing. Like most Hyde Park residents, David's parents were born to sharecroppers in southern Georgia. After David's father got a maintenance job at Fort Gordon, they moved to Augusta and rented a place in South Nellieville, a rough neighborhood not far from Hyde Park. Eventually, the Jacksons bought a plot of land in Hyde Park, near David's aunt. Slowly, over the course of several years, David's father pieced together pasteboard scraps that he collected from Fort Gordon until he built a house on the site where David and I sat that sunny Saturday morning. Finally, when David was six years old, the family moved in. [The kids in school] "used to call it a pasteboard house," he said, adding in a sing-

song voice, "'You live in a pasteboard house,' that's how they teased us." Eventually, the Jacksons rebuilt their house with sturdier materials, but it took many years. "Money wasn't that good," admitted David.

And it's no wonder—the Jacksons raised eight children. To help feed the family, and to stay connected to their farming pasts, the Jacksons cultivated a large, prosperous garden. "We had sweet potatoes this high," gestured David. But David also remembered that as the junkyard grew, the garden's vegetables often came up rotten. "So we quit trying to plant a garden because it ain't going to do nothing," he added.

Life in general got harder for the Jacksons when David was a young teen. His father had moved on from Fort Gordon to take a job at Babcock and Wilcox and then at Georgia Carolina Brickyard. There he "worked off a machine catching bricks" and loading them until his leg got crushed. Although David was being primed for a football scholarship, at fourteen he left school to help support the family. "[My dad] was too proud to get welfare," David told me. "A lot of peoples had welfare, but not us. . . . I figured that if I come out and help my brothers and sisters finish school, later on I might could go back," he said. But David never did finish school. He started out working on a garbage truck. A few years later he moved on to Boral Bricks where, as his father had once done, he stacked and loaded bricks.

Some years later, David married and moved into the family's backyard shed. He, his wife, and his younger sister and brother crowded in there for about a year before moving on. Eventually, David and his wife decided to try their luck in Boston, where "opportunities and jobs were better." David found a job operating a freight elevator for a canning company in Cambridge, but the cost of living was so high that he still "had to struggle still just to maintain." When his dad fell ill, he decided to move back to Augusta. Unfortunately, the day after David arrived, his father passed away. When they were able, David and his wife moved into his father's house, and David went to work for a trucking company. Throughout the 1990s, he continued to work long hours driving trucks up and down the Atlantic seaboard and sometimes across the country.

David would not necessarily say that his dad's death was caused by a life of hard, manual labor or by eating strange garden produce. One thing was certain, though: the fact that they were African American had directed both Jacksons' lives. The jobs David's father could get in the 1950s and 1960s were limited. As a result, he worked in hazardous occupations and got injured, which led David to quit school, thus curtailing his economic opportunities. Racial segregation further limited the Jacksons' housing options. Like the Calloways, the Jacksons lacked the safety nets that many middle-class families enjoy. As a result, both families wound up stuck with unsightly, inhospitable, and downright dangerous industrial neighbors.

From Promised Land to Poisoned Land

There was an inherited love of the land in Jeeter that all his disastrous experiences with farming had failed to take away.

—Erskine Caldwell, *Tobacco Road*

"This is what Hyde Park looked like when I was young." It was early evening in mid-October. I sat in the backseat of the Utley family's midnight blue Buick as we rode through Waynesboro, Georgia, about thirty-five miles south of Augusta. In between rows of Georgia pines, long, dusty fields dotted with light gray puffs of cotton framed either side of the one-lane road. Every so often, we passed a cluster of houses surrounded by wide green lawns. Reverend Charles Utley, fifty-three, a minister and middle school guidance counselor, drove the car and narrated. Next to him sat fifteen-year-old Anthony Utley, a tall, muscular boy who played basketball and sang in his high school choir. Like many kids who spend a good bit of time in the country, Anthony had learned to drive several years earlier. In typical teenager fashion, he frequently remarked on his father's skills behind the wheel. Utley's wife, Brenda, a second-grade teacher, sat next to me on the backseat. We chatted about Demetra, the Utleys' oldest child, who lived in Ohio while she pursued a master's degree in speech therapy. We were on our way to Bible study at the Utleys' "home" church, somewhere on the outskirts of the town (the main intersection of which consisted of two gas stations, a barbecue pit, a hotel, and a McDonald's). As we got closer to the church, Reverend Utley proclaimed that we were in "Utleyland," and the scattered homes we passed in the fading light were those of his relatives.

Like many of Hyde Park's original settlers, Reverend Utley's parents moved their young family to Hyde Park from Waynesboro in the mid-1950s in search of better job opportunities. As we drove through the broad green fields of rural southeast Georgia, Utley continued to compare the landscape to the Hyde Park of his childhood:

> Well, Babcock [an industrial ceramics factory] was there. The power plant was there, but it wasn't as large as it is. The rest of it was just beautiful landscape. Just land, trees, and sage fields. Where the junkyard is, that was the area that we could plant. Nothing there.

Utley's description highlights how, for its original settlers, Hyde Park combined the best of two worlds: surrounded by trees and fields, it appeared bucolic, but, just six miles from downtown Augusta, it was also close to the many job opportunities that the city's growing economy promised. Relocating to Hyde Park, then, gave its residents the chance to take a giant step toward a prosperous future while keeping a toehold in their rural pasts.

The story of Hyde Park begins in the middle part of the twentieth century with the hopes and dreams of its original settlers. These former sharecroppers were the first in their families to own property and build equity to pass on to future generations. However, their dreams came to a crashing halt in the 1990s when residents discovered that nearby factories and hazardous waste–generating facilities had contaminated their land, potentially endangering their health and making their homes almost impossible to sell. For Hyde Park residents, this turn of events resonated with a history of struggle and discrimination.

Clearly, people address environmental problems according to how they conceptualize and understand the "environment." Anthropologists Michael Paolisso and R. Shawn Maloney, for instance, illustrate how Maryland farmers viewed the environment as something that could never be fully understood, quantified, or regulated, whereas environmental professionals believed that science could both diagnose and resolve the region's environmental problems. In both cases, professional identities were the lenses through which people viewed the environment.[1] In Hyde Park, residents' conceptions of the environment and land were shaped by race and class identities. Yet in this case I found that "land" and the "environment," rather than appearing as synonymous terms, were almost antonyms.

Although Hyde Park residents had a long history of interacting with the land as farmers and rural residents, they had traditionally been quite removed from the environment as a movement. Historian Michelle Johnson explains that, for African Americans, the land has generally connoted concrete natural resources. The environment, however, has remained a "nebulous concept" that is not often part of everyday discourse, particularly because it is seen to be a white, middle-class concern.[2] In recent years, however, a formerly "nebulous environment" has become all too real for hundreds of African American communities as they realize that their neighborhoods are contaminated. This chapter illustrates how the environment entered Hyde Park residents' consciousness along with news of contamination, coming to mean something poisonous and discriminatory from which they needed to be saved. Whereas the land had represented the American Dream of ownership and a hope of overcoming racial discrimination, the environment represented the very obstacles that discrimination imposed against achieving that dream. Once they characterized the environment as a site of racism, activists added it to their traditional social justice agendas. They then expanded their definitions of the environment to include all the resources to which they lacked access (i.e., housing, schools, and police protection).

People infuse their physical surroundings with memory, experience, and history, and these interpretations are often a crucial inspiration for collective action. When people's places are threatened, the force of the meanings and identities that they have ascribed to them transforms into powerful forms of activism.[3] In Hyde Park, residents' cultural experiences seeped into the land on which they lived and nurtured a strong community identity, which fed their activism. However, residents' attachments to place also became quite complicated by the end of the 1990s.

One-Eighth of an Acre and a Garden: Land and the Promise of Prosperity

> *Gardens? Yes, we had a pretty garden. Grew turnip greens, collards, cabbage, butter beans, string beans, okra . . .*

When the Utley family moved to their new house in Hyde Park, they found themselves in a burgeoning neighborhood, replete with opportunity. Men could walk to work at Merry Brothers, as well as to Babcock and Wilcox ceramics factory, Piedmont Wood (later, Southern Wood Piedmont), or a new Georgia Power plant. Women could take the bus a few miles into Augusta's "Hill" section to work as domestics. Moving to Augusta, then, offered families a double wage and, with it, a chance for economic mobility. Perhaps even more important than the promise of economic opportunity and dual incomes, however, was the fact that, for fifty dollars, the African American farm laborers of Waynesboro and other nearby rural areas could finally acquire their own piece of land. As Charles Utley explained,

> A lot of families moved in that were [from] the rural areas. Because in the rural areas, you couldn't own the land—you were crop sharers [*sic*]. And that was the reason my family moved here because otherwise they would never have the opportunity to purchase the land. They had to sharecrop it. And they had saved up money to move to this area.

Landowning is a goal for many Americans, but for the Utleys and their neighbors, it had a particularly intense meaning, given the fact that they had come directly from the harsh sharecropping realities detailed in the last chapter.

Moving to a neighborhood like Hyde Park was thus attractive for both practical and symbolic reasons. On the practical side, it was close to a number of industries, and one could buy a substantial portion of land for relatively little money. Of course, one of the primary reasons that the land was cheap was that it was undesirable. Hyde Park is only minutes from Phinizy Swamp, and "the swamp" was a common and well-founded nickname for the neighborhood. But its marshiness also made the neighborhood relatively bucolic. In other words, moving to Hyde Park meant that former farmers did not have to leave their country lifestyles entirely behind. Charles Utley explained why Hyde Park appealed to its early settlers:

> I would say 98 percent of people had a garden. Matter of fact, when we moved, we brought the cow, the hog and everything else. So you didn't lose your country life in the city. Matter of fact, the idea was you could move to the city and keep your country life at the same time.

It is not by chance that Utley prefaced his statement by mentioning how many of his neighbors had gardens.

Gardens played an important role in early Hyde Park life, and they permeate social memories of the neighborhood. Almost unfailingly, when I questioned people about Hyde Park's history, they immediately mentioned gardens. For example, when I asked David Jackson to tell me about his fondest memories of growing up in Hyde Park, he responded,

> When we came out here we didn't have problems with the land a whole lot. My daddy used to have a garden right out there. And we used to raise sweet potatoes, all kinds of good stuff out there. And it was good.

Other residents similarly remembered how, in the 1950s through the 1970s, large and bountiful gardens dotted the neighborhood. As one man who began renting a house in the neighborhood in the early 1960s said,

> Everybody had a garden out here just about I know of. [They] planted a garden because people love to do that.

As this man commented, people love to garden, and gardening is an important part of American history and culture in general. But for the residents of Hyde Park, gardens took on particular significance, given their history as not-too-distant descendants of slaves and as sharecroppers.

Depending on the type of work they did and the hours in which they did it, slaves were sometimes able to cultivate "patches" of vegetables, which they harvested and possibly even marketed or bartered. In Georgia's cotton belt, there is evidence that slaves traveled from rural areas into cities and towns to vend their vegetables.[4] After the Civil War, as black farmers became more transient under sharecropping conditions, it was more difficult to grow gardens. But the ravages of the boll weevil and the Depression made self-sufficiency even more important, and sharecroppers struggled to cultivate even a meager garden. The ability to cultivate and maintain "patches" became a source of satisfaction and pride—by staking a claim and cultivating even a minuscule plot of land, slaves and sharecroppers were able to carve out an aspect of their lives that was not entirely controlled by their oppressors.[5]

Many years later, in Hyde Park, gardens continued to symbolize security, pride, and self-reliance. For example, Arthur Smith recalled how his father would share the produce from his garden with friends and relatives:

> And the crops so good, you would give part of your garden to other people. I remember my aunts coming down and picking greens out of my father's garden and other crops.

Smith's statement implies that having a bountiful garden with enough produce to share was a status symbol. Importantly, Hyde Park residents only occasionally spoke of the grocery money that garden-tending saved them or the sustenance that garden produce provided. More frequently, they emphasized the degree to which a good garden was a source of pride. Smith explained,

> You know the gardens, they were like the pride of the men that lived in this area. Part of feeding your family.[6] I think that's part of the responsibility of manhood that I was taught. You know it was nothing new, certain times of the year, to see my father out plowing his field. He had a little roll plow that had a wheel and he would break the lanes up. . . . It was nothing unusual to see collard greens, tomatoes, okra, squash, watermelon. I remember big plum trees, fruit trees, all types. And each house had their own garden.
>
> MC: Only men worked the gardens?
>
> AS: I'd seen ladies hoe, but it was like, "I'm Arthur Smith and this is my garden and this is my wife." It was a man thing.

Smith's statement ties gardens to male identity. Gardens symbolized not only a man's ability to provide for the family but also his autonomy. Whereas many of Hyde Park's men moved from one factory or laboring job to another, subject to the whims of their employers and the economy, their gardens were stable, productive, and, for the most part, under their control.

For Hyde Park residents, gardens also symbolized racial progress and a chance to salve the wounds of slavery. Because they had grown up sharecropping on former plantation lands, Hyde Park residents lived in particularly close proximity to slavery's ghosts. Possessing and cultivating their own land allowed African Americans finally to own the products of their labor. Community activist Terence Dicks explained,

> I think [southern African Americans] have a point of view that they are after, all of this, entitled to, you know, that American culture that we sacrificed and slaved for and everything. . . . Some people . . . that's the only thing they ever worked for and the only thing they ever had, that's all they got to show for a life of misery, toil, and drudgery.

Owning land thus provided an enduring sense of security that had been denied to African Americans during years of enslavement and exploitation.

In her exploration of return migration, a growing trend among northern African Americans who move back to the rural South, anthropologist Carol Stack similarly argues that for African Americans, owning land promised the kinds of security and liberty that were denied to slaves. She writes,

> The appeal of God's little acre crosses all bounds of race and time, but the urgency could seem shrill for African Americans. If security and liberty were to be found anywhere, wouldn't it be under one's own roof, safe on one's own land?[7]

Stack finds that even several generations after sharecropping, owning land in the South became a metonym for recovering and re-creating history. According to Stack, unlike past migrations of African Americans northward, return migrations are motivated not by economics but by "a powerful blend of motives" that represent "the chance to start something new, to remake the South in a different image."[8] Hyde Park's settlers and the children of its settlers similarly viewed landownership as a chance to remake the past.

Part of this "remaking" included former sharecroppers' chance to build real equity to pass down to future generations. Long-term resident Bernice Jones moved from Waynesboro to Hyde Park in 1954. Jones explained,

> This is a neighborhood where people who were basically farmers and other people who had probably done sharecropping and who had never owned property came together buying property with the idea that they were going to buy something and build homes to have for their children to be raised in.

Real estate does more than provide equity for current and future generations; in many ways, it acts a linchpin for economic and social opportunity. Because land- and homeownership open doors to better education, employment, safety, insurance rates, services, and wealth, barriers to homeownership effectively bar people from social mobility. In marking the first step on the road to prosperity, homeownership often acts as a "key benchmark" in the story of individual African American progress.[9]

Owning land thus gave Hyde Park residents a chance to rewrite a history of racial subordination in a number of ways, all of which converged on the fifty-by-one-hundred-foot plots of swampland that composed Hyde Park. Together, the land and the opportunities it promised became prime ingredients for the American Dream. As Arthur Smith described,

> These houses were bought with sweat and tears. When I say that, I mean sacrifice. It's easy to say, "I'm going to live here and pay rent," but these people wanted something in life. This was my father's first home and my Uncle C.E.'s first home and many other peoples that moved out here. This was the first opportunity for them to live the American Dream, buy a piece of land, and "build me a home and raise my family and then leave it for my children." That's part of the American Dream in this great country.

As Smith indicated, by becoming homeowners, Hyde Park residents achieved new status as Americans.

Homeownership also shored up this status for future generations—another critical part of the American Dream. Anthropologist Constance Perin argues that while renters in the United States are seen as flighty, irresponsible, and itinerant, homeownership constitutes a social category equated with perfected citizenship because it signals having earned the trust of a bank, the sense to invest wisely, and the desire to accrue equity.[10] For Hyde Park residents, owning homes and land represented a chance to join the American mainstream and to claim their place as full, and even ideal, U.S. citizens. In the early 1990s, however, news of contamination stunted the future possibilities that landowning promised. Now homes were unmarketable and gardens were toxic. Thus, the soil in which Hyde Park residents had instilled their American dreams became the very source of their dreams' destruction.

Toxic Soil, Tainted Memories

> *People used to make gardens. I had a good garden. Now they can't make anything. From the contamination, I guess.*

In 1998, memories of gardens were bittersweet. Some residents claimed that gardens became too difficult to grow as industry increasingly

Entering Hyde Park on Dan Bowles Road, 1999. Photo by author.

encroached on Hyde Park in the 1970s and 1980s. David Jackson described how as the junkyard behind his house grew, his father's garden suffered:

> But then when that junkyard moved in there and all this other stuff came in here, the property just went nowhere. You could plant something that's rotten right there before you. When it's trying to grow, it's rotten.

Jackson's account of the demise of his father's garden is especially poignant when contrasted with his earlier description of it in its heyday. For Jackson, quitting the garden marked a significant turning point in his memories of life in Hyde Park.

Other residents faced a similar turning point in 1991, when the University of Georgia (UGA) Cooperative Extension Service for Richmond County tested produce and soil from Hyde Park gardens and found elevated levels of arsenic and chromium. The director of the Extension office later went on record as stating that he would not eat anything from a Hyde Park garden. Subsequent analyses of the test results varied (UGA scientists advised against eating from gardens, while the Richmond County Health Department contended that the food was safe for ingestion); but the results alarmed residents, and almost all decided to let their gardens die.[11] Thus, contemporary memories of gardens were tainted with fear as residents wondered whether and when their health would bear the effects of having eaten garden produce. Ruth Jones said, "When they said not to eat anything, we stopped. We had a beautiful garden." With very few exceptions, the vegetables and fruits that had once been the pride of the neighborhood were left to wither, uneaten and unharvested.[12]

Once contamination was discovered, all kinds of neighborhood memories became similarly associated with health worries. For example, prior to 1970, residents did not receive city water, instead using outdoor pumps connected to underground wells. However, after some reports concluded that both Hyde Park's groundwater and soil were contaminated, residents feared that the water they had pumped had also been filled with toxic chemicals.[13] Although narratives about the backyard pumps usually began with the kind of humor that comes from talking about "the old days," they often turned to anxieties about whether the water had been toxic. Residents also now worried about having attended family barbecues on the SWP site. Wives of SWP workers wondered not only about their husbands' health but also the fact that, back then, they had washed the men's work clothes by boiling them.

Moreover, before gas lines were installed in the neighborhood (also in 1970), Hyde Park residents had cooked and heated their homes with wood-burning stoves. A common chore for neighborhood children was to go into SWP's field, gather leftover creosote-treated wood chips, and take them home to burn in those stoves.[14] Charles Utley remembered,

> We would get the firewood from the chips that they would use to make the wood, and it was easy to burn so we would take that, and we would put it in the heaters and we would heat with that.

However, as Robert Striggles explained,

> You see when we was burning that wood, we didn't know that it was harmful to us. We was burning creosote. We didn't know . . . creosote was a cancer-causing agent.

When residents realized that the wood chips contained creosote, another narrative about Hyde Park's "old days" shifted from emphasizing pride at overcoming hardship to worry over health.

The poisonous chemicals seeping from surrounding factories also had a dramatic effect on Hyde Park residents' community identity. As David Jackson said,

> When they come up with that stuff about pollution and everything else —that killed it. Killed it. Ain't nobody doing nothing to try to upgrade or do nothing. It ain't like that anymore. We're still just downgraded. And people can't have a say or nothing now . . . nobody selling because nobody wanting to buy nothing. They aren't going to pay that kind of money for something contaminated.

Jackson's statement about still being downgraded illustrates the degree to which Hyde Park residents identified with their neighborhood. In some ways, the land had become a metonym, standing for Hyde Park people. In her book on working-class residents of a midwestern city, anthropologist Rhoda Halperin argues that neighborhoods are not just localities in which daily life occurs but places where "threads of identity . . . are woven from memories of work, of family and of neighborhood goings-on."[15] In Hyde Park, owning their own homes and cultivating their own land had once identified Hyde Park residents as people who had achieved a particular goal. After 1990, however, the neighborhood's reputation was "downgraded" as it became associated with contamination.

Similarly, other Augustans now associated Hyde Park with decline and destruction. Accordingly, as they waited for relocation funds to come via their lawsuit or the federal government, many Hyde Park residents stopped investing in their homes, and the neighborhood became

House on Horton Drive, 2003. Photo by author.

increasingly run-down. By 1998, a significant number of houses had peeling paint and overgrown lawns, and ditches were strewn with litter. Thus, as the neighborhood declined physically, residents' pride in it also declined, further "downgrading" their community identity.

Importantly, community identities are never uniform, and members of a community rarely agree on all aspects of an issue. On the one hand, some Hyde Park residents firmly believed that their neighbors should spend only the bare minimum necessary to maintain their homes and keep the area looking clean, since they were working so hard on relocation strategies. In this view, relocation might become a reality at any time—why build up your home if you were about to leave it any day? On the other hand, less hopeful residents insisted that as long as they were stuck in Hyde Park, they wanted to live in the nicest home they could afford. David Jackson, for example, told me,

> If work need to be done around here, I ain't going to wait until somebody say that we going to relocate y'all to do it. My house needs to be a decent place to live so I do that. . . . So people say, "Oh, don't do nothing to your house." What do you mean? I live in that.

Porch sitters, 1999. Photo by Maryl Levine.

In 1998, residents thus faced a conundrum: Should they continue to live in downgraded circumstances and wait for relief, or should they try to make their homes as livable as possible, despite contamination?[16]

Whether or not residents chose to invest money in home improvement, after 1991 Hyde Park *did* decline both physically and in terms of its reputation. Almost all Hyde Park homeowners agreed that selling was virtually impossible once news of contamination had spread throughout the area. Even though studies on whether contamination had affected residents' health had varied widely, property law bound Hyde Park residents to disclose the possibility of contamination. After 1991, most homeowners felt trapped on their potentially poisoned land. Bernice Jones, a retired teacher in her fifties, explained this catch-22:

> People invested in homes, and now they can't sell them for anything like what they paid for them or put into them. Ours will be paid for in November, but where will we buy another comparable house? How will we start over at this age?

Where it had once promised prosperity and the opportunity for self-sufficiency, land in Hyde Park now represented disappointment and the dashing of the American Dream.

These phenomena—tainted memories, the inability to sell a home, the destruction of a dream, and the loss of community—are all typical ways that toxic exposure *socially* contaminates a community, according to environmental psychologist Michael Edelstein. Edelstein finds in these places an "inversion of home" or the "negation of the hopes, dreams and expectations that surround the institution of home in American society."[17] Moreover, Edelstein argues that the closer people are tied to home (i.e., those who work there or who are retired, homebound, and/or members of clustered extended families), the more acute the effects of contamination. Hyde Park shows us two additional factors in this sense of loss: poverty and racial history, both of which tied Hyde Park residents to their land in profound ways.

From slavery to land loss to sharecropping, owning land symbolized overcoming a history of racial exploitation, as well as a stab at a middle-class life. The chance at ownership thus transformed land from a site of racial oppression into a site for racial autonomy, freedom, and belonging. It is no wonder, then, that Hyde Park residents connected the contamination of their land to their history of racial struggle. As the environment flooded Hyde Park residents' consciousness, it rode in on a tide of destructive chemicals and years of discrimination.

"Too Busy Trying to Live"

> *We didn't really think about the environment before. We was just too busy trying to live.*

Mentioning the term "environment" to adults in Hyde Park prompted them to tell of the dust that covered their walls and reappeared as fast as they could wipe it off. They spoke about the toxic release sirens coming from Thermal Ceramics that sometimes blared for eight hours, forcing residents to go to the mall or the movies to escape the noise (let alone the toxic release). They told of how they had permanent tickles in their throats and how their children were never far from their inhalers. They described how their children could not dig in the dirt around their houses or play in the ditches that lined the streets of their neighborhood. For the residents of Hyde Park and for the activists of HAPIC, the environment was not something to be protected from human intervention and conserved for the preservation of wildlife.

Rather, their environment was poisonous, and they needed to be protected from it.

Hyde Park residents had always connected the fact that their neighborhood lacked urban resources such as adequate police protection, litter control, and refuse disposal to its racial composition. In 1968, they formed the Hyde and Aragon Park Improvement Committee as a *civil rights* association to demand city water, sewage, streetlights, and paved roads. Although they had deep connections to the land, with very few exceptions, Hyde Park residents paid no attention to environmentalism or environmental matters per se until they heard that their air, water, and soil might be poisoned. Immediately, residents connected that pollution to racial discrimination. They had diverse conceptions of where this discrimination came from, but for all of them the "environment" was attached to words like "racism" and "injustice."

Even HAPIC's president, Charles Utley, said that before 1990, he "didn't even think about [the environment]." One activist reasoned that the environment was "not really an issue because most people just were not exposed to it." Some residents said that they had heard some things on TV, but most agreed that they never cared about it very much.[18] Blacks in the South are all too aware that they face a long list of institutional discriminations, but they have been slow to realize that the siting of hazardous waste facilities is on that list. Not only have African Americans historically viewed the environment as a white, middle-class concern, but also the employment opportunities that these industries promise often overshadow their detrimental environmental effects. Sociologist Robert Bullard describes how residents of minority areas were hesitant to question corporate and governmental polluters because they did not want "to bite the hand that fed them." In the "black belt," African Americans are more likely to be unskilled, poorly educated, and intimidated by large corporations.[19]

In areas like Augusta, intimidation worked well. One man told me that the reason residents stood by as numerous factories located and expanded in their neighborhood was that "people are afraid of the white man." In another example, Charles Utley explained,

> There was no one to tell the community people, you shouldn't be there. Or to be a watchdog for what chemicals they were producing. It was a way of life that blacks were used to going to work, doing what they were told and not asking why. As a result, damage [was] done to their

> health and the environment. . . . They're [treated like] second-class citizens anyhow.

Although Hyde Park's own history of activism and fighting against institutional racism belies some of these statements, they do highlight the difficulties of fighting environmental racism, even in areas where education and income levels are mixed. First, confronting large, faceless corporations may be more intimidating than opposing local governmental officials. Second, during the civil rights era, activists challenged blatantly racist laws, but environmental racism was an unexpected form of discrimination, and one that was not easily identified or readily recognized.

For many years, Hyde Park residents had resigned themselves to the conditions that their factory neighbors imposed on them as a part of neighborhood life. They grew used to the residue that covered their cars, the oil that often appeared on the surface of ditch water, and the fact that their water sometimes "had an odor to it." Annie Wilson, one of Hyde Park's first residents said,

> That water one year, it was stinking. And we really hadn't paid it that much attention. . . . In Aragon Park, my niece was living over there, one time, that water was so stinking they couldn't take a bath in it.

Some residents, on the other hand, said that they suspected something might be wrong in the neighborhood, but because they felt they did not have much evidence to back them up, they never addressed it. David Jackson remembered,

> My yard used to flood out more than anybody in this whole area because all the water from the junkyard would flow right in my yard. And when it leave, it leave all kinds of grease-filled and black-looking dirt with the oil and stuff that just shot up in here. But for years, we know something wrong, but we don't know exactly what it is.

Some female residents said that they had begun to notice that their neighbors had certain health problems in common. For instance, Johnnie Mae Brown, a sturdy woman with graying hair, was also an active and veteran HAPIC member. Brown quit her job at Fort Gordon long before she was eligible for a pension so she could care for her aging and sickly mother and uncle (who worked at Babcock and Wilcox for twenty years). Brown said,

> We knew that there was something, but we didn't know what it was. . . . The children were breaking out with these different rashes on their skin. A lot of people in the neighborhood died from cancer. I had a sister to die. She was only thirty-two years old.

Brown's experience follows a gendered pattern of environmental justice awareness that is especially common among working-class women. A number of environmental studies scholars posit that women activists are prevalent in the environmental justice movement because they tend to be more closely aware of and affected by family illness.[20] I do not have the space to discuss this issue in detail here, but in my interviews women frequently claimed to have "always" been suspicious of local health problems. Some people also told me that, before Mary Utley herself fell ill with heart trouble in the mid-1980s, she was gathering information on cancer rates in the neighborhood. However, widespread organizing around environmental issues did not begin until the early 1990s.

Until the institution of federal environmental regulations in the 1970s,[21] SWP discharged its residual water into Rocky Creek and Phinizy Swamp, and it burned treated wood waste, producing smoke and fly ash. By 1979, in compliance with new federal regulations (such as the Clean Water Act), SWP had redirected its waste disposal and installed monitoring wells on the plant's property. In 1983, those wells revealed on-site groundwater contamination from wood-preserving chemicals (including creosote, arsenic, chromium, and PCBs). Investigations over the next five years identified two plumes of contamination. One extended approximately two thousand feet from the plant along Winter Road, and another extended east along the old effluent ditch to New Savannah Road. Five years later, SWP decided to close its Augusta plant, partly because the soil and groundwater contamination could not be adequately remediated as long as the plant remained in operation.[22]

Residents of Virginia Subdivision were unsurprised at the plant's closing. At the time, Virginia Subdivision was a mostly white neighborhood right on the border of SWP and across a field from Hyde Park. Several long ditches ran from SWP property through Virginia Subdivision and into Hyde Park. As early as the 1970s, residents of Virginia

Subdivision had begun filing complaints with the Georgia Environmental Protection Division about a foul odor emanating from their drinking water and their backyard wells. They also documented the number of residents with cancer and were alarmed by the results.[23] In response, the Agency for Toxic Substances and Disease Registry (ATSDR) conducted a health consultation in 1987. Its report discouraged residents from using well water and suggested that certain ditches were potentially contaminated.[24] Residents of Virginia Subdivision also joined with several local companies that owned property around SWP and filed a class action suit seeking damages for trespass, nuisance, and neglect. In mid-1990, SWP's parent company settled the lawsuit for approximately $8.6 million.

Although some Hyde Park residents claimed that their water had always tasted suspicious, unlike in Virginia Subdivision, almost all Hyde Park residents agreed that they did not seriously question the conditions of their neighborhood until around 1990. Around that time, a local environmentalist who monitored emissions from Thermal Ceramics claimed to have found PCBs and creosote running through local ditches and encouraged community leaders to keep track of pollution in the area.[25] Even so, many of the Hyde Park residents I interviewed attributed the beginning of their environmental awareness to a large flood in 1990 that swept over Hyde Park and left in its wake a foul-smelling bluish white mud and houses full of corroded furniture. Johnnie Mae Brown remembered the "high water" of 1990:

> Most people in the neighborhood didn't even think about [the environment] until we had that flood. After the flood we knew that something was wrong because that water, everything that the floodwater touched, it was no good no more.

Ollie Jones also recalled,

> That would have been about when that high water was . . . in the nineties. . . . My water was so high it was all in my porch and stuff, and a lady from Channel 12 was interviewing me about the contamination because of the water and all that and I said, yes, it's been like that for quite a while cause the water would come up from the sewers and things. And that was when I first really got into it, you know. . . . [The water] had a funny color and an odor.

Ditch on Golden Rod Street, 2004. Photo by author.

Shortly after the flood, stories about the SWP closing appeared frequently in the local news, and residents also realized that the prison crews working on Hyde Park's ditches had not been around for "quite some time." Indeed, at least a year before the flood, Richmond County officials had halted ditch work in Hyde Park because they were worried about the possibility of contamination.[26]

Also around the time of the flood, HAPIC leaders first heard about the Virginia Subdivision lawsuit. Although the subdivision was adjacent to SWP, whereas Hyde Park was located about a half mile from the factory, it seemed obvious to Hyde Park residents that the ditches lining both sides of their streets carried water directly from SWP through their neighborhood. They quickly linked own smelly drinking water and foul-smelling mud to those ditches. Moreover, the same chemicals found in Virginia Subdivision were discovered within fifteen feet of Clara E. Jenkins Elementary School at Hyde Park's entrance.[27] And some Hyde Park residents alleged that they had seen SWP trucks dumping waste into nearby fields at night.[28] Residents were then incensed when they were excluded from the settlement. Although they had not yet filed any legal actions against SWP, they believed that offering them a settlement would have been "the moral thing to do."[29] In addition, Virginia Subdivision residents had not included Hyde Park in their lawsuit. Because at the time Virginia Subdivision was a mostly white neighborhood and Hyde Park was an entirely African American neighborhood, Hyde Park residents interpreted their exclusion both from the original lawsuit and from the settlement as a clear-cut case of racism.

Just after the Virginia Subdivision settlement, Bill McCracken, a local attorney, approached HAPIC leaders and began organizing a class action lawsuit, asking African American civil rights attorney Harry James to join him. Along with HAPIC leaders, the two lawyers began signing residents on to the suit. Soon former SWP workers agreed to issue statements indicating that they had dumped waste into Hyde Park. McCracken and James also found a medical study produced by SWP that addressed the issue of skin cancer prevalence in dark-skinned individuals compared with that in fair-skinned individuals. However, finding enough proof to satisfy legal standards that SWP deliberately contaminated Hyde Park because its residents are black has never looked promising. The lawsuit, then, makes allegations similar to those in Virginia Subdivision's suit (i.e., trespass, nuisance, and neglect). Connecting their strange, intermittent rashes, asthma, and cancers to environmental contamination led residents to categorize the environment as something dangerous and even deadly, in sharp contrast to their understandings of the land. It also catapulted them into environmental action.

Erin Brockovich Doesn't Live Here Either[30]

I ran into one of the organizers of the Love Canal. And when we compared the Hyde Park story with the Love Canal, theirs was white and ours was black. And that was the biggest issue.

In 1998 and 1999, Hyde Park faced a twofold problem. First, some studies had shown, and residents certainly believed, that toxic chemicals had contaminated the neighborhood and threatened the health of those living in the area. Second, residents lacked the financial resources to move out of Hyde Park, and selling a house was nearly impossible. In addition, lawyers on both sides of the lawsuit continued to file multiple motions, which had the effect of delaying the trial date.[31] Attorneys McCracken and James were working on contingency, so the time they could devote to Hyde Park's lawsuit became increasingly limited as the years wore on. Eight years after filing their lawsuit, most residents had given up hope that they would ever see a legal victory. Despite the tireless efforts and indomitable spirit of HAPIC leaders, the possibility of receiving a governmental remedy for their situation also seemed bleak.

While Hyde Park residents faced the double whammy of being black *and* poor, they overwhelmingly agreed that racism, not classism, was the primary reason for their situation. For example, Charles Utley once told me that Hyde Park's situation had "95 percent to do with race." In numerous interviews, residents stated why they thought their neighborhood had been contaminated and why they had not received any help. Occasionally the initial answer was that Hyde Park was a poor neighborhood. Yet once I encouraged interviewees to be frank, or once my research assistant, Michelle, told them to "go ahead" because I was "okay," they admitted that they felt race was "the biggest part" of the reason. The more residents I interviewed, the clearer it became that racism was the framework from which they viewed their circumstances. (It also became clear that residents were not entirely comfortable blaming their problems on racism in front of a white audience.) Certainly, Hyde Park residents believed that their neighborhood's image as a low-income area contributed to their situation. Yet Virginia Subdivision's settlement confirmed for them that the presence of white (if low-income) people had elicited a series of positive responses from the judicial system, which led to compensation. That Virginia Subdivision residents actually received very little did not mitigate Hyde Park residents' sense of being discriminated against.

For example, I interviewed Ollie Jones, a bus driver for Richmond County Public Schools who had lived in Hyde Park for forty-two years, and his wife, Ruth, a thirty-year Hyde Park resident, in the large, immaculate living room of their trailer.[32] Mr. Jones's carefully ironed khaki shorts and plaid shirt barely hinted at the heat of the day. Mrs. Jones, wearing shorts and a T-shirt, had just gotten off from work cooking lunch at the Jenkins summer school and was a bit breathless from a morning spent with elementary school children on summer vacation. I asked the Joneses to name the reason for the Hyde Park situation, prodding them to be "frank." Ollie Jones responded,

> You want me to be frank with you? If this was a white neighborhood, now I'm being honest, government would've stepped in here and wouldn't have been about two or three words said. Another thing is if this had been a rich neighborhood, it wouldn't have been nobody in

> here, they would have moved them out of here. But most of the people are poor, black people.

Jones went on to compare his neighborhood to Virginia Subdivision:

> If you look at [Virginia Subdivision] over there, for instance, the majority of the peoples over there is white people, right? And Piedmont is connected to this place too, but they were complaining over there, they didn't hesitate, they bought them out.
> RUTH JONES: But they wouldn't do that for us.

For the Joneses, when white people complained about contamination, SWP "bought them out." However, when the black residents of Hyde Park complained, SWP ignored them. Totsie Walker agreed:

> [Being black is] the biggest part. If we was in a mixed-up neighborhood, they would've done something. But you see it wasn't. Now, you see, that's the way I feel about it . . . they just don't care, you see.

Other residents pointed out that, while not everyone in the neighborhood could be considered low-income, everyone in it *was* African American. Johnnie Mae Brown said,

> Yeah, I think we're black, low-income people in the neighborhood. And not everyone is on the poverty line, you know, but we're still in the neighborhood. And I think it's because we're black first of all. That's the most important. We're black, poor people. And they just build anything they want around us and we don't have no say so.

Brown's statement highlights her perception of the powerlessness of blacks in Augusta, regardless of income level, profession, or homeownership.

Many fine books and articles enumerate the multiple and complex problems that contribute to environmental injustice. As a cultural anthropologist adding to this body of literature, I aim to explore in detail exactly how and when African Americans talk about race, especially in the context of the environment. All Hyde Park residents I spoke with believed that some kind of racism was the reason for their situation. Even the few residents who did not accept that their neighborhood was contaminated still believed that no one had responded to Hyde Park's call for help because the people living there were black. [33] For example, Deacon Saulsberry (who denied the presence of contamination) said, "It's hard for the white people to say yes to something and stick to it. . . . They don't want to help. They won't help."

Thus, while all residents attributed their situation to racism, the sources they named ran the gamut between specificity and ambiguity—some mentioned local government, some corporations, some federal policies, and some stereotyping. As described earlier, some of the factories and plants surrounding Hyde Park were there *before* it became a neighborhood. Yet some residents reasoned that it was because of racism that they could only afford to live in the midst of factories in the first place. Others answered that, although SWP and Babcock were there when they moved in, at least four other factories appeared over a thirty-year period. Arthur Smith pointed to racial stereotypes: "I think with Hyde Park and Aragon Park, it was the arrogance again of big companies saying, 'those people are not educated. Those people do not vote.'" Other HAPIC leaders cited more systematic practices of racism, referring to their situation as "environmental apartheid" or "residential holocaust." Charles Utley called it "a form of genocide." For Utley, the "genocide" of environmental racism stems from a deliberate, planned, and systemized racism made up of both corporate greed and discriminatory political institutions. In this view, each aspect of the system facilitates the other—corporate greed initiates environmental racism, and institutional racism perpetuates it. The economic and social roots of racism thus go hand in hand in a tangled and mutually constitutive system of discrimination.

Indeed, environmental racism stems from complex causes. While naysayers argue that siting decisions are market driven, not deliberately racially motivated,[34] even if corporations do use race-neutral criteria when they locate hazardous waste sites, or when a neighborhood grows up around existing hazardous sites, other kinds of institutional discrimination contribute to environmental racism and make it almost impossible for residents to leave and escape contamination. First, the generally white racial makeup of local zoning and planning boards gives African Americans little say in factory or incinerator siting decisions. Second, despite the Fair Housing Act and other civil rights reforms, realtors continue to steer African Americans toward existing "ghettos," and mortgages and home improvement loans are allocated most often to white neighborhoods.[35] Racially differentiated neighborhoods produce uneven property values—that is, black neighborhoods tend to have lower property values. In turn, our education systems, which are financed through property taxes, are vastly unequal. Poorer educations give African Americans less access to the kinds of jobs that would enable them to move out of a contaminated neighborhood. Finally, many of those who could move to a white neighborhood are reluctant to do so because of historical experiences of racism and persistent white antipathy toward integration.[36] Again, all these factors make Hyde Park's situation different from those of middle-class white neighborhoods, whose experiences of contamination are also tragic, but which do not necessarily face the added complications of historical, structural racism in America.

Across the country, African American environmental justice activists similarly insist that, whatever the complex chains of events that have led to their toxic situations, they result from long-standing racism. The overwhelming commonalities of these beliefs point to the continued salience of race for African Americans in the United States, especially for those who bear the brunt of racism on a day-to-day basis.

Adapting the "Environment" as an Organizing Goal

The environment? I don't know, but it used to be awful.

In striking contrast to the lack of environmental concern in Hyde Park prior to 1990, in 1998, 79 percent of residents surveyed admitted to being "very concerned" about the environment.[37] Yet the "environment" that residents referred to was of a specific nature—their dramatic introduction to environmental awareness had fixed residents' associations with it as a dangerous, deadly, and racist entity. Residents thus created a framework for environmental organizing that resonated with their experiences as southern African Americans.

When I asked people about their environmental awareness prior to hearing about contamination, intending to find out whether they had been concerned about issues such as recycling, nature preservation, or endangered species, the answers I received referred only to contamination. As one man said, "The little I thought about it was my Daddy was talking about how his garden wouldn't grow no more. Then I started thinking something was wrong." In all my interviews with non-HAPIC leaders, only one resident, who had once worked at a recycling plant, spoke about the environment in global terms. Overwhelmingly, residents' frame of reference for the environment was restricted to the dangers it presented. For instance, when I asked her to define the environment, Totsie Walker replied,

> I heard lots about [the environment]. Sometimes it would be in the paper, but they'd block it out. . . . Some said arsenic, and some said this. Different things. [Those politicians] talk about it, but they never do anything about it.

Similarly, David Jackson defined the environment by proposing solutions to Hyde Park's specific environmental problems. He said, "Pick up the trash. I would relocate, but you know it's a lot of memories here. My Daddy worked hard to have this and I remember how hard he worked."

In addition to defining the environment as a source of contamination, residents associated it with racism. For example, HAPIC's treasurer, Melvin Stewart explained what most people in Hyde Park thought of the environment: "I don't think it's just chemicals, but a lot of people just think that. Racism is not just chemicals." Although Stewart himself had traveled to a number of conferences on behalf of HAPIC and maintained a broader view of the environment, his immediate association of the words "chemicals" and "racism" with "environment" underscored the way the environment had come to be defined in the Hyde Park community. In turn, these definitions made the environment a locally salient issue.

Expanding Environmental Discourse

> *Environmental justice means doing unto others as you would have them do unto you.*

In the spring of 1999, I sat next to Arthur Smith in the crowded sanctuary of the Unitarian Church in west Augusta. For the first time for each of us, we were attending a meeting of the Augusta–Richmond County Sierra Club. This particular Sierra Club chapter acted more as an outdoor activity and hiking club than a political group and had not been very involved in Hyde Park's struggles.[38] In fact, of the fifty or so people at the meeting, Arthur Smith was the only African American. He leaned forward in his chair and jotted notes on the back of a flyer as we listened to Augusta's recently elected mayor speak about his environmental agenda. The mayor's talk focused on litter control and cleaning trash off roadways. During the question-and-answer period, Sierra Club members pursued his themes and asked about starting a mandatory recycling program, garbage pickup, and bicycle lanes. Next to me, Smith raised his hand. Somewhat startling the other meeting attendees, he stood to ask the mayor to explain his "theory about violence in the inner city."

For Smith, it was perfectly appropriate to address inner-city violence at a Sierra Club meeting because violence *was* an environmental concern for his neighborhood. As noted earlier, in the 1980s and 1990s, employment rates in Hyde Park declined significantly, and drug dealing (mainly crack cocaine)[39] became a fixed local industry. Every two or three months, I would hear that the sheriff's department had staged some kind of "sting operation," and the streets would clear for the next week or two. Eventually, however, the dealers would be back to business as usual.

Drug dealing certainly brought violence with it. Only a few weeks before the Sierra Club meeting, Anthony Ruffin, a neighborhood drug dealer, shot and killed his rival, Michael Young, who lived part-time at his grandmother's house on Golden Rod Street. Seven years earlier, Ruffin and Young had gotten locked in a shoot-out until they were both on the ground and bleeding. After recovering, they made peace. However, one afternoon in early March 1999, Ruffin sneaked up on Young while he was lifting weights in his grandmother's backyard and shot him several times in the back. After the shootout, Arthur Smith, a few other neighborhood activists, and I organized a prayer vigil and candlelight march to "stop the violence." In gathering information for the event, we established that nine people had been murdered in Hyde Park over approximately twenty years, almost always related to drug selling.

The belief that drugs and violence were connected diminished residents' feelings of safety. For example, since crack dealing took root in the 1980s, many senior citizens were afraid to venture out at night, and those residents who could afford them, installed alarm systems in their homes. Yet, aside from a few exceptions, violent incidents mainly arose from domestic disputes. For example, in 1999, one man set his ex-girlfriend's porch on fire (no one was hurt), and another man stabbed his brother in the eye. Other than that, I heard about very few crimes. In fact, at one HAPIC meeting, attendees expressed their pleasure that basketball nets, installed the year before, had remained attached to their hoops, and no attempts were ever made to steal the four computers donated to the community center in the fall of 1998. Even so, most residents constantly worried for their own safety and that of their families and their possessions, especially when dealers were around.[40]

At the same time, drug dealers had grown up among Hyde Park's elders. Describing the period when the dealers were operating outside her house, Johnnie Mae Brown remarked,

> It was frightening; they had guns and all that stuff. But they was very respectful. They called me Miss Johnnie Mae; everybody calls me Miss Johnnie Mae. And they'd never disrespect me. If I'd go out there and tell them to move their car, they'd say, "Man you better move that car for Miss Johnnie Mae." And they would be out there cursing and using profanity, and I'd be out there, "Hey, what's y'all's problem?" And they'd say, "I'm sorry, Miss Johnnie Mae."

Some drug dealers also occasionally became involved in neighborhood activism. For instance, Arthur Smith told me that back in the 1970s, when drugs first established a toehold in the neighborhood, dealers would walk into Christian Fountain Church, place some money in the collection plate, and leave. A few years after I conducted my fieldwork, some of the neighborhood's drug dealers decided to revive the softball league started by Mary Utley. In 1999, one local addict tried to revitalize another of Mary Utley's institutions and hold a fashion show, with local children as models.

Drug dealing, then, was not an acceptable aspect of neighborhood life, but it was part of a much larger and more complicated picture. Many residents realized that drugs provided a singular access to revenue, and while they abhorred the negative effects of dealing on their neighborhood, they also ultimately blamed its stronghold on a lack of choice. Arthur Smith once told me, "I'm not going to go up to [those dealers] and judge them and try to stop them until I can give them a job." Activism focused on drug dealing and violence, then, was as much about education and employment as about "cleaning up the streets."

And efforts to "clean up the streets" were as much about crime as they were literally about trash and litter. In 1998, 65.1 percent of Hyde Park residents surveyed named trash and litter as a very serious problem in the neighborhood.[41] Yet most people believed that no matter how hard they tried to keep the neighborhood clean, the city's inattention to it undermined their efforts. For instance, Johnnie Mae Brown said,

> I like to keep my street clean. If I see paper, I pick it up. And I know, sometimes I hire people to clean around the properties next to me. Clean the ditch. And every day I clean paper out of the ditches because the police, they come through here, but they don't really control our area like they should. The only time they patrol it real often is when we keep calling them about the drugs being sold in front of our property. Then they will send somebody out and they will come around often.

For Brown, the city's refusal to take care of its trash paralleled its refusal to help control criminal activity. Moreover, Hyde Park residents' common concerns about trash and litter, and their efforts to stem it, belie stereotypes that link unkempt neighborhoods to lazy, apathetic, and unkempt neighbors.

All this discussion is intended to convey the point that activism in Hyde Park was broadly conceived—drugs, violence, education, employment, police protection, litter, and pollution were all tied together in an intricate knot bound up in historical discrimination. Thus, not only did residents define the environment as a civil rights issue, but all the civil rights issues HAPIC addressed had also become the "environment." For instance, in 1998–1999, its activities included an after-school environmental education and tutoring program, computer training for adults, showing a series of videos on drug education, conflict management, and teen pregnancy, a prayer vigil and march to stop violence in the community, applying for environmental cleanup grants, and continuing to pursue the class action lawsuit against SWP. From this wide-ranging list, it is apparent that HAPIC continued its original mission to advocate for the civil and social rights of its constituents. As Melvin Stewart said, "HAPIC deals with whatever comes up—crime, drugs, contamination. Contamination is an extension of what [we've] already done."

Although contamination had become HAPIC's main focus, it did not preclude engagement in more traditional organizing activities. Charles Utley explained,

> It's like I tell the kids. The environment is everything you can't see. . . . It is your home. It starts with your home as the world and narrows to your room and to your bed. It includes your messy bed, your messy room.

When HAPIC activists said that they wanted to clean up their environment, they meant that they were working to remediate the ecological damage left by toxic contamination, as well as the social damage left by a legacy of institutional racism. Arthur Smith elucidated,

> [Environmental justice] is health, prosperity. How can you prosper if you're sick? How can you have prosperity if you're deprived of economic values? How can you have the American Dream? . . . I was brought to this house and my father built this house before he met my mother. And I think that goes back to what we were talking about environmental justice and the family. . . . That was a part of the American Dream.

Smith's discussion of environmental justice leads back to his vision of homeownership. For Smith and other environmental justice activists, achieving environmental justice meant the chance to rebuild their lost American Dream and attain the social opportunities that they had historically been denied.

To achieve this goal, HAPIC leaders decided to focus on relocation rather than cleanup. Ideally, they envisioned the relocation of the community as a whole and sought to create a replica of the old Hyde Park.[42] Arthur Smith went on to say,

> I'd like to see everybody move. All the streets out here, re-create them somewhere in the city. . . . And we got the next generation coming up one, two, three. Why not re-create this and then make it the dream I think Mother Utley had for everybody. Make it the dream that Reverend Roundtree had for everybody.[43]

By relocating as an intact community, Hyde Park residents could set themselves back on the road to the American Dream that they had originally envisioned. Achieving environmental justice then meant reestablishing a place that provided the opportunities required for social mobility along with the warmth and friendship of close-knit community life.

Conclusion

Early on, living in Hyde Park meant that you were someone on the move, someone who had overcome hardship and worked hard to buy property to provide your family with a better way of life. Back then, the opportunity to own land had transformed it from a site of the oppression of slavery and sharecropping to a site of promise. After 1990, however, being a Hyde Park resident meant that your dream was endangered and your family's health might be threatened. In short, the environment had permeated and despoiled your dreams, and you were trapped in a neighborhood where no one wanted to live.

Environmental psychologists tell us that upon discovering their toxic conditions, these communities undergo "lifescape shifts," or a process of reinterpreting their health, their past dreams, and their future prospects.[44] But in Hyde Park, such shifts did not generate a full-scale paradigm change. Rather, while residents were well aware that their neighborhood was no longer a place of promise and dream fulfillment, they also continued to associate it with a proud, active past throughout which they had fought for their rights as American citizens. When they found out about environmental degradation, then, Hyde Park activists just added clean air, water, and soil to the long list of resources to which they were already fighting for access.[45]

The ecological circumstances that transformed a social environment of gardens and dream homes into one of toxic contamination and unsellable property were not simply acts of nature; they were the result of a complex combination of economics and politics.[46] Environmental justice activists thus teach us that the environment is both social and ecological, and that everyday categories such as "land" and "environment" may seem transparent but mean very different things for different people. In places like Hyde Park, where people have faced the tyranny of slavery, sharecropping, and Jim Crow laws, and where they continue to face various kinds of institutional and individual discrimination, race is the primary lens through which such definitions are perceived. The environment, then, represents both toxic poisons and the social poison of racial discrimination, and seeking environmental justice means overcoming a long and vicious history of racism in America.

Five

Foot Soldiers

Around Hyde Park, Sharon Palmer was known as "Grandma." All afternoon and on evenings and weekends, children flowed in and out of her house looking for snacks, solace, or both. Sometimes Sharon even intervened in a child-parent dispute, and if things seemed particularly bad, she and her husband, Earl, somehow found some extra room and took a neighboring child in for a while.

It is no wonder the Palmers' non-air-conditioned two-bedroom house often appeared to be bursting at the seams. When all six grandkids were in residence (which sometimes happened for a year at a time), a small "shed room" in the back served as a third bedroom, and clothes hung from bars that were suspended in doorways. A television was usually on, someone was on the phone, and Earl (famous for his cooking) generally had a few pots boiling on the stove.

Sharon was used to crowded homes. One of eight children, she was born in the early 1940s to sharecropping parents in Jenkins, Georgia. She vividly remembered her hardworking childhood:

> *All my life was on the farm. I picked peas, I picked cotton, I picked corn. I even drove a mule. We had cows, we had hogs, we had chickens. Sure did. We would come out of school and go in the field and work. We grew butter beans, collard greens, squash, watermelon, tomatoes, okra, everything in the garden. We did all that—didn't have to go shopping at all really. . . . I stayed home more than I went to school because I had to pick cotton and the dust from the cotton would get in me. I had real bad allergies.*

Sharon and Earl liked to joke that they "met in the field." Actually, they met in elementary school. At the age of nineteen, the childhood friends became sweethearts and married. They moved to Alabama, but soon after, Earl's grandmother fell ill, so they returned to the Augusta area. At first they lived with Sharon's great-uncle in Hyde Park, but then the couple found the small rental house where they lived for the next thirty-eight years. Earl got a job piling bricks at Merry Brothers and then moved on to a job at Federal Paper, a local paper mill.

Meanwhile, the couple had one daughter and then another about a year later. Sharon started volunteering at Mary Utley's day care center. A few years later, Utley helped her get a job with the Economic Opportunity Authority (EOA) cooking and cleaning at different community centers. Finally, Sharon settled at the community center on Golden Rod Street, where she worked with Mary Utley as a community developer, giving small grants to people who needed help paying their rent or power bills. Sharon's community work was not confined to her job, though. In the late 1960s, she joined her neighbors in a battle for water, sewer, and gas lines. Then, in the 1990s, she began to connect reports of contamination in the neighborhood to illnesses in her own family. (Michelle had terrible asthma, and Michelle's youngest daughter had arsenical keratosis.) Throughout the 1990s and on into the next decade, Sharon and her granddaughters could often be seen walking door-to-door, passing out flyers for neighborhood meetings, or helping out at community events.

By that time, though, Sharon and Earl could have used some help of their own. In 1968, Earl slipped and fell at Federal Paper and seriously injured his back, prohibiting him from doing further factory work. For a while, he did some yard work at the community center on Golden Rod Street. But after Mary Utley became too sick to work, these odd jobs dried up. The family then made do with Sharon's salary and Earl's workers' compensation and whatever their daughters could contribute.

The last time I interviewed Sharon and Earl, a new landlord had not only raised their rent but also refused to make much-needed repairs. Sadly, they had moved about a mile away from Hyde Park. The new house was a vast improvement, materially. It had one and a half baths, a washing machine, a full kitchen, and a large living room. Michelle's kids, the only two grandchildren now in residence, each had her own bedroom.

On a sunny spring day in 2003, Sharon, Earl, and I sat in their spacious new living room, catching up and leafing through photo albums. An air conditioner hummed in the background, and, as usual, pungent smells of dinner wafted in from the kitchen. After all their years of surrogate parenting, distributing flyers, and attending meetings, Sharon and Earl admitted that life in their new neighborhood seemed pretty quiet. Even though Sharon still worked at the Utley Center, the neighborhood, especially its kids, certainly missed "Grandma." Sharon told me that one particularly troubled fifteen-year-old who lived with his alcoholic father had camped out in the old house (which remained unrented) for several weeks. One night the boy got drunk, and some neighbors caught him tearfully wielding a can of gold spray paint, decorating the house's facade with the words "Hyde Park."

Long Is the Struggle, Hard Is the Fight

For most of his forty-odd years, Arthur Smith Jr. lived on Walnut Street in the one-bedroom house where he had grown up. Although officially he lived by himself, Smith was rarely alone. A continuous stream of neighborhood kids and adults flowed through his door to raid his ample supply of sodas, chips, cookies, and conversation. Inside, Smith's front room (also his office) was almost as crowded with objects as his porch was with people. Almost every inch of wall space was adorned with certificates, photos, buttons, and other mementos (from the local chapter of the Disabled Veterans, New Hope Community Center, Blacks against Black Crime, HAPIC, and the African American Environmental Justice Action Network, to name a few) that testified to Smith's years as an activist. *Kinte* dolls and angel statues perched on the edge of his desk, guarding either side of two large glass bowls filled with snacks. Behind them, a wood-paneled wall bore several bumper stickers with slogans such as "Stop the Killing," and "Beautify Atlanta." Three other walls displayed posters with quotes from some of Smith's heroes: Martin Luther King Jr., Malcolm X, and Jesus. Propped on a shelf just beyond his threshold sat a gold-plated plaque inscribed to HAPIC's founder, the late Mary L. Utley.[1] Smith's own certificates of merit and honor took less prominent places on his bookshelves, resting against books about African American activism, sociology, and psychology. One small shelf, within an arm's reach of his desk chair, was stacked with a few well-worn and thumbed-through books about environmental justice.

Smith's walls and bookshelves depict his years as a community activist and connect him to a larger history of black activism in the United States. In this chapter I show how the cultural meanings people gave to the land on which they lived translated into a powerful neighborhood identity, which laid the foundations for three decades of tenacious community organizing.

Arthur Smith's front room, 1999. Photo by author.

By the 1990s, many aspects of social movement organizing had changed, and discrimination was not always obvious. Rather than opposing the discriminatory practices of local government as they had in the 1960s, activists were now challenging a powerful multinational corporation, as well as legal discourses and the practice of environmental science itself. Moreover, the social contexts in which HAPIC activists organized had changed: in many situations, race-based organizing was "out" and diversity organizing was "in."

Thus, while HAPIC activists certainly drew upon a civil rights–era legacy, they also expanded that era's traditional modes of organizing, adapting their strategies to meet the demands of contemporary situations. At the same time, no matter how much their strategies and discourses changed, and no matter what kinds of new situations they faced, activists held on to the fundamental, democratic values that formed the foundation of their activism. It was this foundation that sustained them throughout their long organizational history.

"It's Not the Size of the Dog in the Fight, It's the Size of the Fight in the Dog":[2] 1960s Activism

In 1950, along with her husband and seven children, Mary Lou Utley moved to Hyde Park from Waynesboro, Georgia. Over the next two decades, Utley would become an Augusta–Richmond County legend as she steered her Hyde Park neighbors into the oncoming wave of civil rights organizing.[3] Initially, like many of her female neighbors, Utley worked part-time as a domestic in the "Hill," an elite, white district of Augusta. To get to the Hill, the women had to cross Gordon Highway and walk another mile or so up to Turpin Street, where they could catch the bus. With Hyde Park's propensity for flooding, however, these walks were often more like wading. Tired of wet feet and ankles, Utley investigated ways to improve the flooding problem and started asking Richmond County's elected officials to do something about it.

Soon after, George Utley began earning enough money for his wife to give up her part-time housekeeping. All but one of the now eight Utley children were in school or out of the house, so Mary decided to use her newfound spare time to provide child care services to her neighbors. Her brother-in-law, Walter Utley, owned a vacant house and donated it to the cause, and a handful of neighbors gave their time. Most of these women had been helping each other out for some time, meeting on Sunday or Saturday evenings to participate in a Christmas savings club. Domestic jobs provided the women with their own wages, and a limited opportunity to save. At savings club meetings, each member would pool a dollar or so, and then the group divided it equally at Christmas time. The club occasionally also sponsored teas and other fund-raisers to increase the pool.

The savings club signals another way in which the lives of Hyde Park residents contradict certain popular social science theories about African

Americans. In their book *Beyond the Melting Pot* (1963), Nathanial Glazer and Daniel Moynihan famously argued that African Americans did not have strong local attachments, traditions of mutual aid, or patterns of saving money. More recently, scholars have argued that while Yoruban traditions of savings pools carried forward to Caribbean societies,[4] this tradition vanished among African Americans in the United States.[5] More recent historians, however, find that by 1910, in proportion to their populations, black women had developed at least as many voluntary associations as white women. Throughout the twentieth century, the self-help tradition continued to be evident among African American women in the South. Many self-help groups later expanded to include social welfare activities, providing a foundation for activism during the civil rights era.[6] In many ways, Hyde Park's organizing during the civil rights era thus grew out of the informal, church-based networks that its women had already formed, like the Christmas savings club.

Toward the end of the 1960s, Utley wrote a grant application to Saint Alban's Episcopal Church in New York City to initiate transportation and employment programs for seniors and health care and education programs for unwed mothers. In response, Saint Alban's sent a team of its monks to set up shop in the Walnut Street house and assist Utley with her efforts. The monks arrived amid growing anger among Augusta's African Americans over the slow pace of desegregation. At the same time, Mary Utley had not given up on her idea for resolving the flooding problem in her neighborhood. She began to funnel the racial unrest that was simmering throughout Augusta's African American community into a neighborhood-wide effort to lobby the county government for flood control infrastructure, as well as streetlights, paved roads, gas and sewer lines, and running water. For Utley and her neighbors, their lack of infrastructure was a leftover of the Jim Crow system.[7] Fighting for infrastructure, then, was equivalent to fighting against racial segregation. In 1968 (the year the monks came), twenty-two-year-old Charles Utley returned from Vietnam and joined his mother's campaign. He explained the reasoning behind their efforts:

> It was a racist issue in that, see, we had domestic workers, so they would leave the neighborhood, ride the bus, and they would see that in other neighborhoods, these things existed and it was less than three miles from here, everything that we wanted, they had. So it was looked at that it was a racial issue.

According to Utley, witnessing the stark contrasts between Hyde Park and richer white neighborhoods raised residents' consciousness.[8] The nationwide struggle for racial equality further sharpened this awareness and provided ideas for how to channel it into action.[9]

In addition to helping Utley expand her programs, the Saint Alban's monks assisted in incorporating Hyde Park's neighborhood association. By 1969, the Hyde and Aragon Park Improvement Committee had registered as a neighborhood organization with Richmond County and the State of Georgia and had installed Mary Utley as its first president.[10] Ambitiously, the committee's bylaws established a twelve-member board of directors, with each member serving three-year terms. According to its bylaws, HAPIC had a twofold purpose. First, the organization would

> provide channels of communication between segments of the population of Richmond County in order to aid in recognition and definition of problems in the community, to help in organizing community resources, in harmonious body, the people of Hyde and Aragon Park and surrounding areas.

This objective implies that HAPIC organizers were deeply concerned about their perceived isolation from other parts of Richmond County and their lack of access to Richmond County's decision makers. For them, establishing a social movement organization meant embarking on a concerted effort to increase residents' connections to other communities in their county, as well as to county government. This goal became an important theme that recurred throughout HAPIC's history.

Second, the bylaws state that HAPIC would

> assist the people of Hyde and Aragon Park and surrounding areas in obtaining a better standard of living and securing their rights as citizens, regardless of race, creed, color or national origin.

Here, "securing their rights as citizens" refers to voting, which was still an issue in late-1960s Georgia, and to discourses of "citizenship," which were especially being promoted by Korean War veterans.[11] The importance given to "obtaining a better standard of living . . . regardless of race, creed [and] color" underscores how residents formed the organization against the backdrop of the civil rights and Black Power movements. Charles Utley clarified this linkage:

> This was when they were just trying to pull together and voice their complaints at being segregated. It was an issue of being denied privileges of those things. . . . [It was] a civil rights thing that it was mostly geared towards.

Obtaining infrastructure was, then, part of the fight against a Jim Crow legacy.

Between 1968 and 1970, Mary Utley and a cadre of neighborhood activists doggedly attended Richmond County Commission meetings and badgered council members until they acknowledged Hyde Park's problems. Gussie Coleman, who moved to Hyde Park in 1957 from Hephzibah, Georgia, recalled,

> One time we went down to the courthouse and they was talking about what they gonna do. So I held up my finger and I said, "we are citizens of Hyde Park, we are citizens of Augusta, and if we don't pay our taxes out here, you're gonna take our homes away from us. We're supposed to have as much as the [other] peoples out here, water, sewage, and all that."

As mentioned earlier, for the residents of Hyde Park, homeownership (and the taxes that come with it) signaled citizenship. Coleman's emphasis on her and her neighbors' rights as citizens again links their struggle to those of the civil rights era, which strove to realize full citizenship rights for African Americans.[12]

The residents' tenacity paid off. Two years after HAPIC's incorporation, Richmond County's Department of Public Works sent a truck to pave Walnut Street. Shortly thereafter, the county ran water, sewer, and gas lines out to Hyde Park, paved the rest of its streets, and installed streetlights. It also began a drainage improvement program and sent prison crews to dig ditches along both sides of every street in Hyde Park. Mary and George Utley donated a small plot of land they owned on Golden Rod Street, and the county transplanted an old army barracks from Fort Gordon to house the new Hyde Park Community Center. Officially, HAPIC owned the land on which the center sat, and the county recreation department owned the building. The county paid the utilities on the building, but HAPIC was responsible for its management.

A handful of residents credited Ed McIntyre, then an African American county commissioner, with finally securing infrastructure for the neighborhood. One resident suggested that credit lay with the monks and other "white folks" who helped Utley, but an overwhelming majority believed that Mary Utley's skillful leadership and her ability to organize her neighbors were the main reasons for their success in the late 1960s. Indeed, residents revered their deceased leader. After her death in 1990, Richmond County officials renamed the Hyde Park Community Center the Mary L. Utley Community Center. When I arrived there in 1998, Utley's name stood out in bright white letters against the center's redbrick face. Underneath her name, more white letters read, "She was a strong leader, a warm and loyal human being." Septuagenarian Annie Walker echoed the sentiments of many when she told me,

> [Mary Utley] was a Christian. . . . She was a good helper now. Sure was. Would go out and ask for things. Would go out and ask for help. I know it's been harder for the older people since she passed. Ain't no think about it.

Significantly, many residents emphasized Utley's skill at motivating her neighbors and keeping the community together. Richard Johnson, a meat packer who moved to Hyde Park from Waynesboro, Georgia, in the early 1960s and has lived there ever since, recalled the contagion of Utley's spirit:

> Mrs. Utley was a tough lady. She was determined not to give up. She wanted better for her children and better for all of the kids out here in the swamp. . . . Everybody was willing to do what they could do. Always have been. Aragon Park and Hyde Park always have worked together. Something needs [doing], we did the best we can.

Johnson's recollection stresses how credit for past successes was shared among all neighbors.

These interpretations contrasted with those of local academics. For example, historian James Carter attributed the installation of sewer lines to the passing of a Richmond County ordinance stating that every residence had to have indoor plumbing, via septic tanks or sewer lines. Ernestine Thompson, a professor of sociology at Augusta State University, agreed with the residents who believed that the presence of white monks had persuaded the mostly white county council to establish water lines. According to Thompson,

> They came in and lived in Hyde Park, and that's when you got these things in the paper, "These holy white men need water, they need sewage," and that's when they got it.

Such academic interpretations exemplify the "resource mobilization" approach to social movement study, which argues that without the connections and financial resources of white liberal groups, the mobilization of many civil rights protests would not have been possible.[13]

But my purpose here is not to posit a "correct" interpretation of how Hyde Park activists won their water and sewer lines. Rather, I wish to point out that in the 1960s, they were not an entirely isolated neighborhood (although they stated that they often felt that way) but had established useful networks with outsiders. This pattern of working with others then carried forward into their more contemporary organizing strategies. More important, most residents mentioned neither monks, legislators, nor legislation when describing how they won their infrastructure. They almost unanimously agreed that it came directly from Mary Utley's leadership and their own forbearance. These narratives about protests in the early 1970s are as important as the events themselves. Casting neighborhood success as having been internally (rather than externally) generated affirmed residents' belief in their ability to effect changes. In addition, residents' narratives about Hyde Park's historic battle for infrastructure resonated with the more general story of civil rights in the United States. In the late 1960s, they achieved some crucial gains in equality, thanks to a charismatic leader who was now deceased. Residents' ideas about their historical power as a community bolstered a powerful collective identity that carried forth into the decades that followed and sustained collective action in the years to come.

"Hold On, Hyde Park": 1970s and 1980s

"Well, we was basically just trying to keep the neighborhood together," HAPIC activist Robert Striggles answered when I asked him about HAPIC's work in the 1970s and 1980s. Although few serious crises emerged during these two decades, Hyde Park activists continued to work on community problems and engage in political protest as needed. For example, until the mid-1980s (when Mary Utley became too sick to work), the community center housed a five-day-a-week preschool that included a cook. It also ran "Operation Mainstream," which hired out low-income workers, an after-school tutorial program, and a program that placed seniors in part-time jobs and drove them to doctor appointments. In addition, in the summer HAPIC took local children on field trips to such places as the Columbia Zoo in South Carolina, a Procter and Gamble factory, and the Sunshine Bakery, and local agencies like the health department organized seminars for teenagers. HAPIC also encouraged Hyde Park and Aragon Park residents to hold small parties and events at the center, and having funeral repasts there became an established tradition.

The Economic Opportunity Authority, a state-financed nonprofit community service organization, funded most of the center's new programs, and in the mid-1970s, the EOA hired Utley to oversee them. Besides being able to earn a salary for her community work, Utley now could draw on EOA's resources to expand Hyde Park's services. This transformation from activist to nonprofit employee follows a common post–civil rights era pattern. After the 1960s, many former movement leaders were hired to work in War on Poverty programs. Some argue that this professionalization of activists weakened radical movement goals and sometimes generated tensions between former grassroots leaders and their constituents.[14] In Mary Utley's case, however, employment in a state-run agency diminished neither her political activism nor perceptions of her loyalty to the community.

In 1974, for example, Utley led a new neighborhood battle when the Richmond County Commission threatened to close Clara E. Jenkins Elementary School due to its dwindling class sizes.[15] After several months of protest, HAPIC activists secured an agreement to keep Jenkins open until at least the year 2000. Johnnie Mae Brown remembered,

> They tried to close [Jenkins] once before. And Mr. Utley's mother, Mrs. Mary Utley, she fought to keep, she and a lot of other folk in the neighborhood, fought to keep Clara E. Jenkins open then.

Once again, Brown first attributes credit for successful activism to Mary Utley; then, on second thought, she spreads it to "a lot of other folk in the neighborhood."

The Jenkins fight also shows that, although they addressed only one political issue in the 1970s and 1980s, Hyde Park residents sustained their political consciousness, and when issues came up, HAPIC had a ready cadre of people to tackle them. Here we clearly see how concentrating on specific, organized events misses crucial aspects and continuities of social action.[16] Rather, local organizations nurture indigenous consciousness and networks over long periods, becoming powerful resources for subsequent mobilizations.[17]

This consciousness forms the basis for what I refer to as "quiescent politics," that is, the retention of political awareness, leadership, and organizational skills in practical consciousness during seemingly dormant periods. In Hyde Park, Utley's leadership and the neighborhood's protest successes bolstered a collective identity that cast residents as able to overcome local power structures with grassroots agency. Continuing to work together further sustained this identity throughout the organization's seemingly less active phases. Thus, in the 1990s, when activists became aware of environmental problems, they were ready to draw on shared social memories, values, and identities to mobilize into collective action.

Throughout the 1990s, HAPIC also continued many of its regular activities at the community center, setting up a neighborhood watch and organizing community clean-ups, senior citizen programs, and after-school tutoring and summer youth programs. While saving themselves from toxic chemicals became a priority, activists never lost sight of other neighborhood needs. At the same time, the fight against contamination eventually shaped the ways in which they understood all their struggles.

"Look to the Hills from Which Cometh My Strength": 1991–1992 (Lawsuit 1)

In 1991, Hyde Park residents faced an awesome battle, particularly in legal terms. The existence of environmental justice and civil rights laws seemed to provide avenues and channels of support to activists like those in Augusta, but in fact residents had little hope of winning their claims. Filing a class action lawsuit against SWP and its parent companies, ITT Rayonier and ITT Corp. (SWP et al.), was among residents' first strategies, but proving racism would have been a nearly impossible task.[18]

Although many grassroots environmental justice groups initially file race-based lawsuits under the equal protection clause of the Fourteenth Amendment to the U.S. Constitution, almost none win these suits. Since

1976, the U.S. Supreme Court has defined "race discrimination" as "intentional or purposeful conduct on the basis of race, or at least some consciousness of race as a factor motivating conduct."[19] Intent must also be attached to individual actors. Thus, to prove that a community is contaminated because of its racial or ethnic makeup, a plaintiff must provide evidence that a specific person or group of people deliberately caused the contamination as a race-conscious act. Because most contaminations happen over long periods of time and for a variety of race-related reasons, it is almost impossible to prove intention. In the well-known environmental justice case *R.I.S.E. v. Kay,* for example, over a twenty-year period, the Virginia County Board of Supervisors approved three landfills, all located within one mile of neighborhoods that were at least 95 percent African American. The population of the county, however, was 50 percent African American and 50 percent white. Finally, when the county approved a fourth landfill to be located in a 64 percent African American neighborhood, community members filed a lawsuit to stop it. However, they lost their case when a federal district court ruled that the community members could not prove that any one individual or group of individuals had *intended* to discriminate against them.[20]

Legal experts generally recommend that environmental justice groups take alternative courses of legal action, such as filing more procedural claims under Title VI of the Civil Rights Act.[21] These types of claims charge local governments with violating the procedures outlined in the National Environmental Policy Act (NEPA) or by Bill Clinton's 1994 environmental justice executive order, both of which stipulate that individuals are not excluded from participating in, or denied the benefits of, federal programs based on their race, color, national origin, age, sex, disability, or religion. NEPA in particular sets out guidelines for public participation in siting decisions. For example, local governments are expected to hold a public hearing and solicit input from community members on the scope of an environmental impact statement, to provide adequate notification for those hearings, and to give communities time to respond to environmental impact statements once they are completed. But often local governments do not follow these guidelines, or they actively try to circumvent them. An environmental justice community might then reverse a landfill permit on the grounds that it did not receive notice about a public hearing, or that the hearing was held during daytime work hours when no community members could attend. Title VI claims, however, have also had low rates of success, partly because NEPA guidelines are vague, and local governing boards can easily argue that they followed the letter of the law, if not its spirit.[22]

Another reason for the high failure rate of environmental justice lawsuits is that environmental law cases rely on statistical data, which, for minority groups, do not adequately account for their experiences. Indeed, conventional environmental science methodologies for assessing the hazards of certain chemicals present a major hurdle for U.S. environmental justice groups like HAPIC. Environmental toxicology is a statistical and probabilistic, or inexact, science (and it can always be contested with different sets of statistics), and many scholars agree that contemporary science is actually incapable of completely resolving the exact level at which a chemical will pose a risk to humans.[23]

There are many reasons for this inexactness. Briefly, environmental toxicology experiments are based primarily on animal studies and then extrapolated to humans, despite the fact that animals and humans often react very differently to chemicals. Lab rodents are also bred to be genetically similar, which makes them even less comparable to genetically and geographically diverse people. Moreover, most studies extrapolate from animals to humans using healthy white male workers as a standard.[24] The ATSDR study on Hyde Park, for instance, analyzed fish samples from a nearby fishing pond and estimated the likelihood that a seventy-kilogram adult who consumed eighteen grams of fish a day for more than one year would become ill from that fish. The study found that the fish posed no danger.[25] However, it is well known that people of color (including children, the elderly, and sick people) consume closer to twenty to twenty-four grams of fish per day.[26] In other words, standard comparison techniques fail to provide information on the range of ways in which women, children, elderly, or already sick people—far more susceptible subgroups—might react to a chemical.

Environmental hazards are also studied under "normal" conditions in laboratories rather than as they are released or disposed of. As a result, scientists often base their assessments of risk on conditions that are actually very different from those a particular community is experiencing. Finally, high-dose studies concentrate on immediate responses to exposures, but many diseases have long latency periods, and their link to harmful chemicals may not become evident for many years. For example, birth defects have especially delayed onsets, and many cancers do not show up for twenty to forty years. Although scientists might be able to establish cause-and-effect relationships between one chemical and one disease under controlled conditions, the chances of establishing definitive cause-and-effect relationships in the real world are slim.[27]

Once risk assessment procedures leave the lab, things become even more uncertain and complicated. When asked to evaluate environmental exposures and risk in a particular community, environmental scientists isolate data and focus on the effects of one chemical at one time. However, many polluted communities (especially those that are low-income) contain dozens of different chemicals from different factories, not to mention particles emitted from cars, trucks, and trains.[28] For instance, many Hyde Park residents spent years working at various chemical-producing plants in the neighborhood. Many also were exposed to chemicals throughout their lives by burning creosote-coated wood in their stoves and drinking well water. Furthermore, the neighborhood is located between two sets of railroad tracks and adjacent to a highway. Hyde Park residents were thus exposed to multiple chemicals at multiple times and in multiple places.

Even if the manifold exposures that Hyde Park residents encountered were included in risk assessments, scientists know very little about the cumulative effects of such exposures. In its "Guidelines for the Health Risk Assessment of Chemical Mixtures," the EPA admits that its formulas contain a great deal of uncertainty, especially because they do not account for synergism, or how chemicals interact with one another.[29] Making things even more complex, the illnesses that communities like Hyde Park complain of, such as developmental disorders, asthma, and circulatory and respiratory problems, generally result from a range of genetic, environmental, and social factors. Indeed, in some cases, an illness may not be directly related to a particular chemical, but it is exacerbated by (or it exacerbates) toxic exposure. For example, two common health problems in low-income African American communities are hypertension and diabetes. Hypertension can lead to kidney disease, and diabetes creates metabolic impairments. Both situations then inhibit the body's ability to process toxic exposures.[30]

Hyde Park residents soon discovered that, given its probabilistic and inexact nature, environmental science is also susceptible to the interpretations and biases of individual scientists.[31] Indeed, much of the environmental risk assessment process relies on the decisions of individual scientists, who might be subject to a host of biases. For example, a recent study conducted by several geographers applied six toxic indices to the same area and found that they yielded widely varying results. The study concluded that "comparing findings across studies and developing generalizations about levels of risk to low-income and minority populations is difficult, if not impossible."[32] Moreover, risk assessments are usually prepared when a business, agency, or corporation seeks to initiate or continue a hazardous activity. These entities hire the risk assessment agency to conduct their evaluations, making assessors highly susceptible to pressure.[33] Finally, following standardized testing protocols is quite expensive, and only well-funded environmental consultants can adhere to all of them, making alternative and less well-funded studies vulnerable to contestation in court. Despite these inherent biases, our social valuations of science persistently overestimate its abilities to provide an objective resolution to certain issues, particularly environmental risk. And legal decisions still rely on the "objectivity" of environmental science as a critical source of evidence.

Throughout the 1990s, HAPIC activists adapted their organizing strategies to such impediments. First, their lawsuit against SWP et al. charged the company with trespassing, lowering residents' property values, and diminishing their quality of life through contamination. Second, in many of their initial public presentations, rather than emphasize their racial identities, activists represented themselves as homeowners whose quality of life had been ruined by local polluters. For the first time, HAPIC joined forces with residents from Virginia Subdivision to find extralegal remedies. Many Virginia Subdivision residents claimed that they had not received enough money from their settlement to move, yet its conditions required them to refrain from any more lawsuits.[34] A coalition strategy reflected changes in a post–civil rights era. In the late twentieth century, "multicultural" or "diversity" organizing became a celebrated and politically desirable mode of collective action. By joining with Virginia Subdivision, HAPIC broadened its power base and made itself more appealing to a general public.[35]

The union between Hyde Park and Virginia Subdivision, however, proved to be somewhat tricky. HAPIC had a longer history of activism and neighborhood organizing. As a result HAPIC activists often took the lead, sometimes appearing alone but speaking for Virginia Subdivision residents and presenting the two neighborhoods as united in that they were all homeowners. At the same time, HAPIC activists maintained a racialized view of their situation; thus, they alternated between race-evasive and race-specific discourses. As anthropologist Ruth Cardoso argues, the discourses used to describe activist identities are "a play of mirrors, through which the grass-roots groups construct their self-image so that it reflects their dialogue with different interlocutors."[36] HAPIC activists were savvy to the fact that certain contexts called for certain kinds of self-images, and their various organizing identities reflected that knowledge.

After filing their 1991 lawsuit and joining with Virginia Subdivision, HAPIC activists recalled successful strategies from the 1960s and 1970s and began to pressure local lawmakers to help them relocate. For example, in March of that year, they appeared en masse at a Richmond County Commission meeting. HAPIC secretary Eunice Jordan, who taught first grade at Jenkins Elementary and had a master's degree in education, demanded that the commissioners divulge more information about contamination. In addition, she presented (both orally and in writing) a list of neighborhood complaints, ranging from potholes to drug trafficking to road congestion around the Goldberg Brothers scrap metal yard. In her statements, Jordan reminded commissioners, "[Hyde Park residents] are not beggars, but we are professionals. We are teachers, welders, retired military, and others who work and send our kids to college." Later, HAPIC president Charles Utley explained to reporters that their goal that day was to acquire information, as well as to "get our image cleaned up."[37] In striving to "clean up their image," activists worked to present a more realistic picture of their neighborhood and to challenge stereotypes of urban black neighborhoods as full of nonworking and unambitious people.

Richmond County's black commissioners responded to HAPIC activists sympathetically (a pattern that occurred on local, state, and federal levels), but they did not make up a majority on the commission. They also had very little power over local corporations and were willing to go only so far in pursuing municipal solutions to Hyde Park's costly problems. While in the 1960s HAPIC activists concentrated on local policymakers, who had the power to grant them the municipal services they needed, in the 1990s they struggled against large corporations with far-reaching and complex webs of corporate power. For instance, in March 1991, SWP hired an environmental consulting firm to conduct a risk assessment that would determine whether the dioxins found on-site presented any health risks. Two scientists working for that firm later charged that SWP officials had manipulated the dioxin data in order to portray the site as less hazardous.[38] While this charge may have bolstered HAPIC's case, it also demonstrated the formidability of the committee's new foe and the need for innovative strategies.

In the first few years after filing the 1991 lawsuit, Hyde Park underwent a battery of tests on its air, water, and soil. These tests produced widely disparate data. It was at that point that activists began to view science as a tool that could be used in multiple ways, with multiple results. Or, as Charles Utley explained, "If I set up the test and the test instrument, I can pretty well dictate the outcome." For example, early in 1991, the ATSDR, a division of the U.S. Department of Health and Human Services, concluded that Hyde Park soil did not pose a significant health risk for residents.[39] HAPIC activists contacted the University of Georgia's Extension Service and asked it to perform another set of tests. This time, test results revealed extremely elevated levels of arsenic, chromium, and other heavy metals that could also be found on the SWP site.[40]

By finding their own testing agency to counter the ATSDR's tests, HAPIC activists also recognized the subjectivity of scientific conclusions. At the same time, like many grassroots environmental groups, they recognized that science was essential to making their legal case; thus, they contested science with more science, finding alternative studies to support their claims. However, in most cases (and certainly in Hyde Park) statistical battles only confuse the issue and do little to advance a community's case. Eventually, many grassroots groups give up on statistical studies and instead assert their own pragmatic and observable experiences.[41]

Facts may be fungible, but as HAPIC activists often insisted, their everyday observations told them that "something is wrong." For instance, more than one-third of Hyde Park residents reported having family members with hazardous waste–related health problems, and time and again they enumerated their experiences of poor community health.[42] David Kimbrough, a vivacious man of forty-five, grew up in Hyde Park and continues to live there with his wife and daughters, near his mother and younger brother. Kimbrough related how his father and brother, both of whom worked at SWP, died at early ages. Kimbrough said, "Matter of fact, [my father] died right there on Southern Wood Piedmont grounds. He was walking, and he just fell dead." Kimbrough also remembered speaking with a doctor who remarked that a just-buried Hyde Park resident was the tenth person he had seen die from almost identical symptoms. According to Kimbrough, his own daughters suffered from hyperventilation, dizziness, and fainting spells until he and his wife stopped allowing them to play outdoors. Many other residents report the strange and untimely deaths of family members. For example, Johnnie Mae Brown's sister died of cancer at the age of thirty-two, Totsie Walker's adopted son (who also worked at SWP) died before he reached forty, Robert Striggles's brother died of cancer in his fifties, and Ollie Jones died of a heart attack at the age of fifty-five.

HAPIC activists were thus caught in an increasingly familiar conundrum: science may be biased, and "facts" can be deconstructed, but was there not some definable truth out there that would validate their claims? Such questions plague environmental activists and academics alike:[43] How are we to prove that global warming exists, if we also argue that all science is biased and subjective? While these questions may not be entirely answerable, one approach to them is to explore who benefits from which science. Not all "facts" have the same power—those that might serve the interests of poor people and people of color are often disregarded in favor of those that serve corporate and political interests. We might, then, examine whose scientific facts tend to be considered valid. Whose voices are heard and whose are disregarded?

In 1991, the voices of HAPIC activists were finally heard by Richmond County's new health commissioner, Dr. Frank Rumph, a tall, soft-spoken African American physician. Rumph recognized the uncertainties of science and also the ways in which those uncertainties can be weighted to favor those in power. In other words, as one activist put it, Rumph was a "friend to the community" from the get-go. Robert Striggles even argued that getting Dr. Rumph involved was "the most important" event of the early 1990s. One of Rumph's first tasks was to publicize the results of a 1991 mortality survey that found death rates in the SWP area to be five times higher than in comparable communities. Although the results of this study were later disputed, it inspired a great deal of fear both in community members and in Rumph himself. Indeed, the new commissioner soon went out on several limbs for Hyde Park residents. First, he publicly admitted that he would not live in Hyde Park, Aragon Park, or Virginia Subdivision.[44] Second, he persuaded the county to install signs around the area warning children that playing in ditches could be hazardous to their health. Finally, Rumph received scientific data with a cautious and wary eye and, on more than one occasion, took the community's side in disputing test results. For example, he publicly stated that he had seen an oily substance in Hyde Park's ditches that smelled like creosol.[45] Rumph also openly maintained that science could be inexact and even alluded to how it could be race and class biased. In a 1998 interview, he said, "Often, people who are passing opinions on the community don't live in the community."[46]

Various health officials bore out Rumph's statement by telling Hyde Park residents that health problems were either "a figment of [their] imagination" or the result of poor diet and exercise habits. Here again, HAPIC is like many other environmental justice groups whose health problems are blamed on their lifestyles (i.e., smoking, eating the wrong foods. and not exercising) rather than on environmental toxins.[47] Certainly there are correlations between income levels and smoking, diet, and exercise, but these factors provide too easy an "out" for health officials. Blaming the victim allows officials to not even examine the question of toxic contamination, which in turn might entail placing responsibility for community members' poor health squarely on the shoulders of polluters and local governments.

Rumph's support bolstered the resolve of HAPIC residents, who stepped up their efforts to be relocated. This time, they carried their complaints to state and national levels. In January 1992, Hyde Park and Virginia Subdivision activists crowded into a town meeting with state legislators to request emergency funds to help them move. Here again, they found that black legislators offered the most sympathetic ear. African American state representative Ron Brown, for instance, convinced the governor to appoint a thirteen-member task force to look into the SWP-area situation. Brown also oversaw the appointment of task force members, assigning Dr. Rumph as chair and including HAPIC secretary Eunice Jordan, vice president Beatrice Holiday, and two residents from Virginia Subdivision. The task force, known as the Governor's Task Force on the Long-Term Health Care Needs for Southern Wood Piedmont Residents, also included several doctors from across the state and a member of the EPD.

Activists also continued to fight locally for financial relief and to lobby for public support by presenting themselves as a deserving neighborhood. For instance, two weeks after the January town meeting, they appeared again before the Richmond County Commission. Arguing that news of contamination had rendered their properties essentially valueless, they requested property reassessments and tax waivers. Unlike the last time she had spoken to commissioners and emphasized Hyde Park residents' middle-class lifestyles, this time Eunice Jordan characterized her neighborhood as economically "poor." "But," she added, "as far as our neighborhood family, we are not poor."[48] Here Jordan strategically emphasizes her neighborhood's image as "poor" to underscore its need for tax waivers. At the same time, by referring to her neighborhood as rich in "family," she persists in disrupting stereotypes that characterize black neighborhoods as disorganized, pathological, and undeserving of public support.

Throughout their first year of environmental justice organizing, HAPIC activists employed a threefold strategy. First, they de-emphasized their race and presented themselves as homeowners to link themselves with, and elicit sympathy from, a wider population. In so doing, they also subtly refuted stereotypes that conflated race and class. Second, they developed critiques of the legal system and environmental science and adapted their strategies accordingly. Third, they extended their opposition beyond the local level and began to lobby state and federal agencies for help.

"Sick and Tired of Being Sick and Tired": 1992–1993 (Lawsuit 2)

HAPIC filed its second lawsuit in 1993 under the name Southern Wood Association for Medical Problems (SWAMP). The lawsuit asked ITT Corp., ITT Rayonier, and SWP to compensate residents for regular health screenings. It also marked the culmination of activists' growing fears about their health and their growing familiarity with environmental toxicology. Now they stepped up their activism over health issues and set their sights further up the political ladder in a new set of strategies designed to make their case known to powerful politicians.

Several major studies led up to the lawsuit's filing. In 1992, a dermatologist observed a high number of SWP area residents with arsenical keratosis.[49] That same year, a neuropsychologist found a high degree of neurological abnormalities among Hyde Park residents, which he believed could be attributed to exposure to toxic chemicals.[50] But the ATSDR issued another health assessment, based on data collected between 1987 and 1992, which found no clear or significant relationship between contaminants and poor health.

HAPIC activists contested the ATSDR report, pointing out that it was based only on soil samples and ignored groundwater, air, fruits, and vegetables.[51] In addition, activists contended that testers had used plastic spoons to collect dirt rather than boring deeply into the ground. David Jackson said,

> They sent out some people to do that testing out here and they scoop[ed] a little bit of dirt with spoons on the ground. Hey, I done put dirt on top of dirt trying to get rid of the floods and things we been having out here for years.

According to Jackson, testers had actually sampled new soil that residents had imported and put over their old, contaminated soil.

Here again, Hyde Park is not unique. Issues such as how deeply to bore and exactly where and what to test are another major hurdle for environmental justice groups. When agencies like the EPA agree to collect soil samples to decide whether a neighborhood is contaminated, they determine where and how to collect the samples. Generally, they do not solicit community input, and the valuable knowledge that residents have about their own neighborhoods is left untapped.[52] In addition, residents are well aware that many cancers take decades to develop. Because they feel left out of the process and believe that life-or-death decisions are being made by people who are not invested in their communities, residents are often angry and resentful even before an agency announces its findings.

Responding to the community outcry, the EPA stepped into the Hyde Park situation and announced that it would conduct a $1.2 million study in conjunction with the ATSDR to examine "conclusively" SWP area soil, air, and water and associated health risks. Upon reviewing the testing protocol, however, HAPIC attorneys believed it had serious "omissions and technical deficiencies."[53] HAPIC board members advised residents not to allow testers onto their properties. EPA officials quickly scheduled a community meeting. Eventually, they quelled community opposition, and testing proceeded. This challenge to EPA methods again illustrates HAPIC activists' increasing sophistication with scientific knowledge. It also signals their ever-heightening suspicions of governmental officials.

HAPIC publications reflected this growing anger and suspicion. Whereas HAPIC newsletters from 1992 urged readers to not give up and to "look to the hills from which cometh my strength," flyers in

1993 read, "Aren't you tired of being sick and tired?" and encouraged residents to "be mad and let them know it." Around this time, activists formed SWAMP and filed their second lawsuit. Mounting frustrations and mistrust of state and federal agencies also led HAPIC activists to set their sights on higher levels of political authority. In 1993, they contacted their new congresswoman, Cynthia McKinney, an African American liberal Democrat. In the spring of that year (several months after she took office), McKinney came to the Mary Utley Community Center to listen to residents' complaints. Later she helped arrange a meeting between HAPIC leaders and EPA officials in Washington, D.C. In 1994, McKinney wrote a letter on behalf of Hyde Park and Virginia Subdivision to the Centers for Disease Control (CDC) in Atlanta, requesting that the CDC release information from ATSDR studies and increase its attention to the "double insult" that residents incurred first from illness and second from expenses related to illness. (Unfortunately, McKinney had a limited effect on the Hyde Park situation.)

In December 1993, the ATSDR released an addendum to its 1992 health consultation. The addendum reiterated the registry's initial conclusions that the community was "probably not being currently exposed to contamination from the site"; thus, SWP did not constitute a *current* public health hazard. However, the addendum also stated that prior to 1993 people had been exposed to unsafe levels of site contaminants in ditches, Rocky Creek, Phinizy Swamp, and residential areas adjacent to the site on Nixon and Winter roads. Moreover, children playing on contaminated soils surrounding the Southern Wood Piedmont site could have ingested the soil, causing skin rashes and possibly skin cancer, and past exposures to contaminated air during plant operations and current exposures during cleanup efforts could cause health effects such as anemia and lung irritations. Accordingly, the addendum suggested posting warning signs around ditches, restricting access to Rocky Creek and Phinizy Swamp, and possible biomonitoring for residents exposed to contaminated soil or well water.[54]

Growing ever more frustrated, HAPIC activists looked even further up the political ladder and wrote a letter to President Clinton in an "urgent plea" for his "help and intervention." Signed by HAPIC's board on behalf of Hyde Park, Aragon Park, and Virginia Subdivision, the letter describes all three communities collectively: "We are a low socioeconomic group, minority neighborhoods, who are and have been poisoned by chemicals." Indeed, by this time, many white families had found ways to move out of Virginia Subdivision, and reduced property values primarily attracted poor African Americans. Whereas in 1986 only six black families lived in Virginia Subdivision,[55] by the mid-1990s, it was known as a mostly black neighborhood, so partnering with Virginia Subdivision now required less of an emphasis on class alone. Writing to President Clinton also reflects the beliefs of many African Americans in Augusta, who referred to him as "the first black president."[56] By appealing to him as "minorities," activists hoped to increase Clinton's sympathy for their situation. HAPIC received only a form letter in response to its request, but coincidentally, the following month Clinton signed his executive order on environmental justice. More significantly for this discussion, the letter to the president signals that HAPIC activists were now more consistently asserting their racial identities in public.

Several months after writing the letter to Clinton, SWP area residents received more confounding and disturbing news. In the late winter of 1994, they gathered at Jenkins Elementary School to hear the ATSDR's response to the EPA study. The study took ninety-three soil samples throughout Hyde Park and fourteen groundwater samples from temporary wells that were installed in the neighborhood.[57] The scientists then isolated the chemicals they found, measured them, and compared them to EPA/ATSDR standards for toxicity. On the one hand, the EPA found highly elevated levels of lead and PCBs in ditches near the Goldberg Brothers scrap metal yard. In addition, two locations indicated elevated levels of arsenic, lead, or dioxins. Chromium and lead exceeding EPA comparison values were detected in eleven groundwater wells. On the other hand, upon reviewing new sampling data from 1993, the ATSDR stated that currently, there were not enough instances of any chemicals to constitute an "urgent health hazard." Moreover, it concluded that the soil did not constitute a significant threat to residents' health unless they "inadvertently ingested it on a daily basis for many years."[58] The agencies' puzzling findings illustrate some of the inexactness that plagues environmental science.

For Hyde Park residents, such contradictions fueled a collective fury at the meeting. One angry man from Aragon Park hefted a chair and threw it onto the elementary school stage. Activists refused to accept the EPA's conclusions or logic. Activist Richard Johnson remembered,

> [The results] didn't make much sense to me. . . . Some of them was saying that it wasn't contaminated out here, that it was just certain parts

> and the ditches was maybe and stuff like that. You know, water come[s] out of the ditches and whatever's in there [is] going to come out when the water come[s] out.

Initially, residents had pinned high hopes on the state's potential to rescue them, despite their extant suspicion of government agencies. One man said, "When we first heard about the EPA study, we thought the cavalry was coming in." Now, however, they saw the state agency as complicit in blocking their pursuit of environmental justice. Hyde Park resident Ollie Jones remarked,

> I don't think [the results] were true. I think they were hiding something. The government didn't want to get the people upset or they didn't want to pay up.

In addition, while "ingesting soil" is a "term of art" commonly mentioned in environmental risk assessments, it is also a common stereotype in Georgia that African Americans eat dirt. Thus, it was immediately understood to be a racial slur.

In response, HAPIC investigated the testing agency contracted by the EPA and discovered that SWP had contracted extensively with this agency in the past. Although they could not *prove* that its past affiliations had actually biased the agency's results, HAPIC activists believed that they had found a clear conflict of interest. They no longer trusted that the EPA was necessarily a disinterested party and began to suspect it of racial bias. Now they left off emphasizing a class-based identity and called directly on their past, transforming their struggle into a full-on civil rights fight.

"Environmental Racism Is Alive and Well in Augusta": 1994–1995 (Lawsuit 3)

In March 1994, shortly after the tumultuous meeting at Jenkins Elementary School, HAPIC activists stood on a narrow street lined with pear trees and redbrick row houses. Across from them, the white marble facade of the eight-story Richmond County courthouse shadowed downtown Augusta. In front of the brightly painted door to Congresswoman McKinney's office, HAPIC activists announced to newspaper reporters and television news crews their intention to file a civil rights complaint against the EPA. In addition to their discovery about the EPA's consulting agency, activists revealed that they had evidence that the Georgia EPD had endangered their community by concealing data on the release of hazardous substances.[59] They also claimed that the state attorney general's office never enforced environmental cleanup plans for either SWP or Goldberg Brothers scrap metal yard. According to the compliant, all these oversights and injustices were due to Hyde Park's racial makeup. Obviously, in charging both U.S. and Georgia governmental agencies with racism, the press conference was a highly notable event in and of itself. But I wish to emphasize that the press conference marked a turning point in Hyde Park's environmental battle: now activists were consistently and publicly linking their environmental problems to their race.

Although from the beginning, HAPIC activists framed the contamination of their neighborhood as a case of racism, they had initially avoided using racialized language in their protests. However, as they encountered health officials' stereotypes, the shady practices of SWP and now the EPA, and a lack of substantial support from local, state, or federal governments, they shifted to a racial basis for their claims. In addition to their concerns over SWP, residents were also recognizing the degree to which all the industries surrounding them potentially endangered their health. They began to refer to their neighborhood as a "toxic donut" and to assert that they faced "environmental genocide." Ignited by the most recent testing debacle, residents no longer couched their struggle in race-evasive terms. Rather, they kicked off a new set of organizing strategies drawn directly from civil rights era movements.

To assist them in planning and implementing this new phase of their fight, HAPIC activists contacted veteran civil rights era leaders, Connie Tucker of the Atlanta-based Southern Organizing Committee (SOC) and Damu Smith (of both SOC and Greenpeace). Over the past few years, both organizers had been key figures in building the national environmental justice movement. Importantly, a strategy of calling in regional reinforcement (also common in civil rights struggles) was designed to bolster HAPIC's strength, not compensate for its weakness. As Robert Striggles explained,

> See when [SOC] come down, they thought they were coming down to organize a group from the ground up. But really, we was already orga-

nized. They were advisers more than anything else. We'd already organized for paved streets, water, everything else out there in Hyde Park. We was already organized.

Striggles's statement illustrates how memories of successful activist efforts in the late 1960s provided the foundation for an organizing infrastructure in the neighborhood, as well as an "organized" community identity.

In embarking on a civil rights–oriented course of action, HAPIC activists mobilized a widespread base of support, both in the neighborhood and in Augusta's black community. First, they organized a boycott of ITT and ITT Rayonier subsidiaries, particularly Sheraton Hotels.[60] Next, they decided to hold a series of marches during the week of the 1994 Masters Tournament—the one time of year that Augusta generated worldwide attention. In gathering support for their protest, activists drew on traditional civil rights networks. For example, they urged local pastors with a letter headed, "Environmental Racism Is Alive and Well in Augusta":

> Churches have historically played a pivotal role in our struggles for justice. This new civil rights battlefield to stop the environmental genocide of our people, and other people of color and poor people urgently needs the support of all our churches.

Appealing to pastors by harkening back to the past proved successful: two local black churches opened their doors for HAPIC prayer/pep rallies, and several black pastors agreed to hold signs and march in the Masters week protests.[61] In addition, a number of Augusta's veteran civil rights organizers attended HAPIC's protest planning meetings, assisting with contacting press and county commissioners and organizing marchers.

On a sunny spring morning in April 1994, as long lines of golf fans snaked around Berkmans Road waiting to view the Masters Tournament, Hyde Park protesters greeted them with picket signs. Wearing long white T-shirts with "Hyde Park" spelled out in bright red block letters, children stretched their arms high and held up signs reading, "We Will Be Your Future, So Save Us ITT-SWP." Adults' placards proclaimed, "ITT You Can Run but You Can't Hyde," "Environmental Injustice in Hyde Park," and "We're Being Killed for the Almighty $$

by ITT-SWP." The judge overseeing the HAPIC case was accused of bias in bold black letters that said, "Shame on You Judge Bowen." Across town, at Augusta's Bush Field Regional Airport, approximately fourteen protesters met incoming tournament-goers.[62] A handful of Virginia Subdivision residents also participated in the protests, but mainly the words "Hyde Park" and "environmental injustice" and the faces of Hyde Park residents headlined local TV news coverage and were splashed over the front pages of Augusta's newspapers. Each test result that denied Hyde Park residents' experiences and fears propelled them toward the Masters week demonstrations. When residents' tensions and frustrations finally exploded, they burst into a decidedly racialized universe.

In one sense, the Masters week protest failed to move HAPIC toward its concrete goal of relocation. The legal process continued to be slow. Because test results were indeterminate and it was nearly impossible to point a definitive finger in just one direction, federal and state agencies promised more of the "same-old, same-old." For example, a few months after the protests, the ATSDR announced that it would conduct more testing of toxins and "attempt to reconstruct the doses long-term residents could have received."[63] Eventually, the ATSDR also produced the amendment quoted earlier.

While the EPD promised to do something to stem the contamination coming from Goldberg Brothers scrap metal yard, its director, Harold Reheis, insisted, "The danger to people's health is essentially not there."[64] Meanwhile, activists kept up their pressure on the EPA, the EPD, and the ATSDR, demanding more meetings and public forums. As activist Reverend Bobby Truitt remembered, during this period, HAPIC "practically harassed the EPA." Although the meetings eventually led the ATSDR to amend its findings (as described later in this chapter), it did not significantly change its conclusions, and community enthusiasm began to diminish. By November 1994, a flyer advertising yet another public meeting with the EPD asked residents to "attend this meeting, so that they [the enemy] will know that we are still fighting them."

In another sense, however, the Masters week protest succeeded. Many Hyde Park residents and activists I interviewed cited it as HAPIC's most successful endeavor. Arthur Smith explained,

> I think the protest of 1994 was a major breaking point for Hyde Park. Again, it goes back to the sixties movements. We went out with the same artillery that Dr. King used—signs, troops, messengers standing

on the corner walking up the street saying we're not afraid to go to jail cause this time we're going to jail for a reason.

For activists and residents alike, the Masters week marches connected them to a proud history of black protest. Thus, activists used the protests to seek concrete results, as well to assert a powerful identity. For example, to television crews that appeared at one of the marches, Charles Utley explained, "What we're trying to let them know now is that we're not going to just step aside and hope it goes away." Utley's statement illustrates that concrete outcomes aside, protesting itself was a means for residents to take power into their own hands, if only temporarily.

In publicly naming SWP, ITT, and Judge Bowen as enemies, Hyde Park residents contested the legitimacy of existing power structures by embarrassing, exposing, and calling attention to those who wished to subordinate them. If we consider the protest an "unmasking [of] dominant codes, a different way of perceiving and naming the world,"[65] then we can see how to HAPIC members, the Masters week protest was a success simply because it happened, and it happened very much in the public eye. For example, SOC director, Connie Tucker said, "[The protest] really did work. It brought a lot of press and a lot of attention to that community." For HAPIC activists, the significance of that week lay in the fact that "the whole world" witnessed not only their plight but also their power to protest it.

Like residents' memories of how they acquired neighborhood infrastructure, reactions to the Masters week protest tell us that definitions of a social movement's "success" should be looked at from a local perspective. Moreover, these interpretations reflect the systems of value and belief that produce collective actions.[66] Together, Hyde Park residents saw their protest as part of a long process, stretching all the way back to the civil rights era. For them, the value of the protests was linked to their pride in civil rights era struggles, as well as the importance they placed on generating widespread attention, countering externally imposed characterizations of their community as disorganized and powerless. For a few days, they were able to show "the world" that they were organized, powerful, and strong. This perspective on the protest was then woven into the fabric of activists' collective memories, and it underscored their future actions.

Indeed, the protest was part of a simultaneously backward-looking and forward-moving process. The attention it generated was the first step in drawing HAPIC activists into the national environmental justice movement. Charles Utley remembered that after the protest,

> [the United] Church of Christ with Charles Lee, they came in. . . . Then they did a seminar, so it just started moving from our little group on out.

That summer, Utley spoke at a Greenpeace-hosted reception for U.S. journalists who covered environmental issues. In September, Arthur Smith traveled to Atlanta to appear before U.S. EPA officials and call attention to a national problem of dioxin poisoning. Although activity within Augusta's boundaries declined (a second Masters week protest in 1995 is notable for the relative lack of press coverage and the decline in participation), HAPIC had permanently widened its focus and broadened its networks. It had also publicly articulated links between race and the environment. Thus, it entered a new phase of environmental organizing.

"He Serves Those Who Wait": 1996–1998

By early 1996, Hyde Park's situation appeared stagnant. The 1993 SWAMP lawsuit had been dismissed because it included a number of Virginia Subdivision residents who had already signed releases agreeing not to take further legal action against SWP. According to attorney Bill McCracken, with so many plaintiffs disqualified from the suit, it no longer constituted a class action. The U.S. EPA's Office of Civil Rights eventually found no wrongdoing in response to the civil rights complaint. Only the 1991 case was still pending, and a fairly steady stream of motions continued to delay the trial date.

The following year, amid much controversy, the Governor's Task Force also concluded that it could not find sufficient evidence to support claims of contamination. Importantly, an appendix to the report that is almost as long as the report itself contained ten rebuttals from people on the task force. One of these, attorney Jeff Bowman, argues that the report fails to include data on pentachlorophenol and chlorinated dioxin levels in groundwater. If studied, Bowman contends, these data would indicate that Virginia Subdivision residents (at least) faced a much greater risk from their groundwater than task force conclusions indicated. Both the dermatologist and the neuropsychologist whose study results were questioned by the report also wrote rebuttals that defended their findings. Finally, Dr. Rumph himself included a rebuttal in which he noted that he had "heard about too much illness and observed too much death" in Hyde Park and Virginia Subdivision. He called for a full medical evaluation of the residents of these neighborhoods and for their relocation while the studies were conducted.

One afternoon in the summer of 1999, I sat in the spacious blue-carpeted office of HAPIC treasurer Melvin Stewart and discussed this period of time. Although he did not live in Hyde Park, Stewart had once worked for the EOA, where he befriended Mary Utley. Later, at Paine College, he worked with Charles Utley. His friendship with the Utley family drew Stewart, who had always been a community activist, into Hyde Park's struggle in the early 1990s, and a few years later he became HAPIC's treasurer. Over the air conditioner's hum, Stewart suggested that delaying the lawsuit might have been SWP's strategy all along:

> They have the funds to prolong these cases as long as they want to. They're not paying any money out, so they can just wait for people to die. I think they think the longer they make people wait, the less active they will be. That's really what's happened as far as the lawsuit's concerned. We don't even hear from the attorneys no more.

Other activists I spoke with took their suspicions even further and accused SWP of actually buying off HAPIC's attorneys, as well as one or two local leaders who had suddenly dropped out of the fight. Such allegations have never been substantiated. What matters here is that HAPIC activists believed that companies would go to great lengths in order to not pay them.

Resigning themselves to the fact that their lawsuit might never pan out, HAPIC leaders looked to longer-term solutions to their problems. In 1996, Dr. Rumph won a $170,000 grant from the Centers for Disease Control to conduct a three-year survey of SWP area residents' health problems.[67] Also in 1996, Utley (again following a pattern of post–civil rights activists)[68] ran for county commissioner. Although he did not win the election, Utley claimed that the publicity he generated for Hyde Park made the experience a victory of sorts. Reflecting on the campaign, he said, "I didn't win, but I had the opportunity to let everybody know that the environmental issues were here and were going to stay." Utley's interpretation of the election again points to the importance activists placed on drawing public attention to their problems.

HAPIC leaders also continued to examine other sources of contamination in the neighborhood. For years, SWP had claimed that Goldberg Brothers scrap metal yard and Thermal Ceramics were responsible for many of the contaminants found in Hyde Park's soil. In 1997, HAPIC activists mounted a brief but successful attack on the scrap yard. Over the years, the business had expanded until its piles of tires, car parts, and rusting metal spread into the backyards that lined the west side of Walnut Street. Water flowing from the ditches around the scrap yard had often looked suspicious, and certain test results (combined with activists' increased knowledge about toxic chemicals) alerted residents to the possibility of serious contamination in and around it. Earlier that year, Arthur Smith attended a regional environmental justice conference, where some of his colleagues advised him to contact the EPA's Emergency Response Team (ERT). Smith returned from the conference and quickly organized a neighborhood-wide phone and letter-writing campaign. Eventually, the EPA reappeared in Hyde Park. This time, ERT scientists determined that Hattie Elam's yard on Walnut Street contained dangerous levels of PCBs and warranted a $100,000 cleanup, financed by the U.S. EPA.[69] However, the ERT did not find significant enough contamination in other yards around the Goldbergs site to merit similar cleanups, although they did order Goldbergs' owner to erect a concrete retaining wall designed to stem the flow of chemicals into the yards of Walnut Street residents.

In the middle to late 1990s, HAPIC leaders thus began to take advantage of the new networks that emerged from the various environmental justice conferences they attended and to explore new organizing options. Melvin Stewart explained,

> Mr. Utley and I sort of took it upon ourselves to begin exploring some other alternatives because we didn't want to just sit around. We had mixed feelings about the lawsuit and some other internal things that were going on.

According to Stewart, growing skepticism about the lawsuit and mistrust of the local political and judicial system prompted HAPIC leaders to shift their strategies. In the process, and due to the "internal things"

Stewart mentions (which I address in the next section), three leaders emerged.[70]

Charles Utley, Melvin Stewart, and Arthur Smith now led HAPIC into its new phase of organizing, which entailed traveling to a great number of conferences (once again, reflecting the importance of "public relations" as a strategy). Indeed, the conferences had a major effect on the life of HAPIC over the next several years. Between 1995 and 1997, for example, HAPIC leaders attended the South-wide Environmental Justice Network development meeting in Atlanta; the Transportation, Environmental Justice and Social Equity Conference in Chicago; and the People of Color and Disenfranchised Communities Environmental Health Summit in Waveland, Mississippi. Through such conferences, HAPIC activists were present at milestone events in the course of the national environmental justice movement.

In 1996, Utley and Stewart attended a "Brownfields 1996" conference in Pittsburgh that set them on a new course toward relocation. Brownfields are defined as abandoned or underused properties in urban areas.[71] In 1995, the Clinton administration's EPA developed the Brownfields Program to investigate these sites, clean them if necessary, and redevelop them as environmentally friendly businesses that would also employ local residents. Because many Brownfield sites are located in poor urban areas, initiatives are intended to stimulate economic development, as well as to address environmental problems. Although the program initially did not provide residential relocation,[72] it gave HAPIC leaders new hope and redirected their environmental strategy. Stewart explained,

> We went to several conferences that were dead ends. One day we got this letter from the Brownfields Initiative. They were having a conference in Pittsburgh, and I said to Charles, "Let's go to it." . . . By the time we left Pittsburgh, we knew that we was on to something. We weren't sure what, but we said this is the best lead that we've had, including the lawsuit, as far as getting this relocation.

Stewart, Utley, and Smith reasoned that securing a Brownfields Pilot Assessment Grant would give them another foot in the door to federal funding. Their faith in the seriousness of their contamination was strong, and they believed that new rounds of tests from a different source within the EPA would reveal significant contamination issues, eventually leading to relocation.

In seeking their Brownfields grant, however, HAPIC activists faced familiar roadblocks that smacked of racial discrimination. For instance, it is customary for municipal governments to apply for Brownfields grants, so in 1998 the activists asked Augusta's mayor to apply on behalf of Hyde Park. Up for reelection, the mayor agreed, but upon reviewing the application later, it seemed obvious that he assigned a staff member who knew nothing about the Brownfields program to prepare it. The resulting product was subpar and later rejected. With their usual forbearance, HAPIC leaders decided to reapply in 1999. This time, they capitalized on their environmental justice networks and asked John Rosenthall, an African American attorney and founder of Howard University's Urban Environment Institute, to assist them. Rosenthall, whom they had met at the 1996 Brownfield conference, had an excellent track record of getting grants and was quite familiar with the Brownfields program. Along with Utley, Stewart, and another local African American leader (Paine College president Shirley Lewis), Rosenthall met with newly elected mayor Bob Young. Young agreed that if the community wrote an application, he would sign it. However, it soon became obvious that Young's administration had little faith in HAPIC's abilities. When news of the application reached local media, for instance, the county administrator appeared on television, stating that Brownfields grants were highly competitive and the county did not expect to receive one.

Nonetheless, over several days in the early spring of 1999, Rosenthall, Utley, Stewart, and I put our heads together and wrote a competitive grant. We chose the former Goldberg Brothers scrap metal yard as our Brownfield site, hoping that once the EPA realized how contaminated just that area was, it would find a way to relocate all of Hyde Park's residents. In June 1999, the EPA announced that Augusta was one of fifty-seven cities awarded a 1999 Brownfields Pilot Assessment grant. The grant mainly funded environmental investigations at the scrap yard (with the ultimate goal of analyzing the cost of redevelopment), but it also included a community involvement component that established a computer resource center in Hyde Park and several community seminars.

Such accomplishments stemmed directly from HAPIC leaders' more general abilities to adapt their strategies as necessary: when their legal tactics and prospects for direct governmental assistance looked bleak, they capitalized on their new state- and national-level networks and looked for alternative tactics. HAPIC leaders' ideas about how to redirect the organization's path were not always readily accepted by HAPIC members, however.

"We've Come This Far by Faith": 1999

Late afternoon on a mid-August Thursday in 1999, Arthur Smith and I corralled a few Hyde Park children into helping us heft and unfold fifty or so metal chairs and set them up in the main meeting room at the Mary Utley Community Center. Most Hyde Park kids spent their mornings at the Jenkins summer school and ate lunch there through the state's "feed-a-kid" program. Afterward, many of them wandered up Golden Rod Street to the Utley Center for its two-hour afternoon program and after that stuck around to color, play kickball, or type on computers. This particular afternoon Smith and I were grateful for the extra hands. We were expecting a larger-than-usual turnout for that night's community meeting, where we planned to answer questions about the Brownfield grant. News of the grant had just appeared in the *Augusta Chronicle* and on television, and ensuing rumors had already shot from house to house and up and down Hyde Park's streets. Many residents believed that the grant meant relocation, and emotions were running high.

Whereas in the early 1990s, almost all residents considered themselves HAPIC members and regularly attended its meetings, now only a core group attended consistently. All residents were eligible to participate in HAPIC programs, and many sent their children to the community center and participated in cleanups or giveaways. Yet most people relied on word of mouth rather than meeting attendance for information. A 1998 neighborhood survey shows how support for HAPIC had waned by that time: merely 10 percent of residents interviewed said they were "very satisfied" with HAPIC in general, and only 14 percent said they were "very satisfied" with HAPIC leadership.[73] One resident confided to me that most people felt that HAPIC leaders "took too long" to accomplish things. Indeed, despite all the news conferences, lawsuits, protests, and meetings of the early 1990s, little measurable progress had been made toward either a neighborhood-wide cleanup or relocation. While they still considered events like the Masters week protest successful, residents were weary of waiting for relief. The Brownfields grant was the first positive news about the contamination situation they had heard in a long time.

Right around the meeting's 6:00 P.M. start time, people began to file in and fill the rows of chairs. Some came straight from work in hospital uniforms or coveralls. Others wore sport shirts and slacks or brightly colored shorts and blouses. Older women (who, as usual, outnumbered older men) covered their gray hair with wide-brimmed hats and leaned on canes. In his characteristic suit and tie, Reverend Utley began the meeting with a prayer asking God to "guide our minds and guide our thoughts so that whatever we do, we do it for the community." After the prayer, Utley announced that, in addition to the Brownfields grant, HAPIC had received a small Community Development Block Grant to install signs at the entrance and exit to the neighborhood. Utley began to explain that HAPIC had a few hundred dollars left over from the second grant, but he quickly interrupted himself to say,

> We don't have the money. Let me say that now. The city has the money. We were thinking of using it for beautification of the neighborhood. But you decide what you want to do with it.

Residents met Utley's comments with silence and chair shifting. His statement and prayer alluded to a set of rumors that had taken hold of the neighborhood over the past few years and which were partly responsible for diminishing meeting attendance.

Since branching out and networking with other organizations, HAPIC leaders had applied for and received a number of grants. They often announced these grants at community meetings and explained their plans to use them in various HAPIC programs. Those who did not attend the meetings heard about the grants second- or thirdhand. Grant funds (which generally ranged between $5,000 and $20,000) seemed like a lot of money to Hyde Park residents, and many believed that HAPIC leaders siphoned off of them for personal use. Each new car, piece of expensive clothing, or home improvement was noted and discussed.

The rumors exemplified two significant changes in social movement organizing in Hyde Park since the civil rights era. First, many African American neighborhood groups increased their affiliations with various institutions, solidifying organizational hierarchies as community organizations "depended less on the political mobilization of residents than on the tactical support of local political elites."[74] Because only a few people were invited to attend out-of-town conferences, the activities of HAPIC's leaders often seemed mysterious. In addition, in obtaining funding from various agencies, leaders took on new commitments. In other words, as they applied for and received grants from the EPA and various foundations, HAPIC leaders found that they needed to comply with program priorities, guidelines, and expectations. These tasks sometimes distracted them from the immediate needs of their constituents, leading constituents to question their accountability. Because full community participation was no longer necessary for all of HAPIC's activities, and because HAPIC's leaders' commitments were now more divided than they had been in the past, they found themselves in a tenuous position vis-à-vis other HAPIC members.[75]

Second, the rumors reflected recent changes in Hyde Park's class structure. In the 1950s, Hyde Park residents had more or less similar economic and occupational statuses. They had all been sharecroppers and were all making a new start in Augusta. After the passing of civil rights legislation in the 1950s and 1960s, however, some people were able to secure better jobs, and the neighborhood became stratified in terms of class. For example, Charles Utley's father, George, came to Hyde Park as a former sharecropper and then worked for Babcock and Wilcox factory. After a series of factory jobs, the senior Utley was able to land a position with the Georgia Highway Department; Mary Utley could then quit her job as a housekeeper and focus on community needs. Charles Utley, a Vietnam veteran, attended Paine College on the GI Bill. A good number of the Utleys' neighbors also received college and graduate educations, but many other people in Hyde Park had to quit school at an early age to supplement family incomes and had never returned, limiting their chances for better-paying jobs. Others got injured on the job at a relatively early age or quit work to help an ailing family member and remained on fixed incomes for the rest of their lives.

By 1999, HAPIC's leaders tended to have higher incomes and education levels than their constituents. Those who did not, like Arthur Smith, who was a disabled veteran and lived on a fixed income, were still able to travel extensively as a result of their role in HAPIC and were perceived to be "middle-class."[76] The activism of Hyde Park's more educated and prosperous residents counters certain popular theories that blame many of the problems of impoverished neighborhoods on the desertion of the black middle class. However, this involvement did not negate the potential for interclass antagonism or, at least, jealousy. Indeed, those residents who most insisted on the rumors were usually the least financially well off. While African American class stratification (and interclass antagonism) has always existed, it was perhaps particularly pronounced in Hyde Park, where class stratification happened rapidly, over the course of one generation.

Mistrust of movement leaders has historical antecedents, and in some ways it is also a by-product of the civil rights era. As far back as the nineteenth century, elite African Americans worked to "uplift" their impoverished brothers and sisters; however, these efforts often turned out to be patriarchal and pejorative. Some argue that this tradition carried forward into civil rights movements. Political scientist Adolph Reed states, "The [civil rights] movement had begun as a result of frustrations within the black elite, and it ended with the achievement of autonomy and mobility among those elements."[77] This history may have shaped and underscored HAPIC members' suspicions of their leaders, particularly if those leaders seemed to represent middle-class lifestyles.

Elsewhere, I have argued that due to this legacy and the general ambiguity of class categories (especially in the United States), environmental justice activists have defined class according to their values concerning accountability. In other words, the category "middle class" was often counterposed to the category "grassroots." You did not have to have a blue-collar job or a particular income to be considered grassroots. Rather, you had to value and be accountable to the needs and concerns of working- and lower-class people. These categories also shifted according to various contexts. In regional and national environmental justice contexts, for example, Charles Utley, was generally considered to be grassroots. Among his own constituents, however, his accountability, or "grassroots-ness," was sometimes called into question.[78]

Shortly after the 1999 meeting began, Charles Utley asked the audience for permission to leave to attend a work-related function. Apologizing, he said, "I do need to keep my job." Meeting attendants responded with silence and some throat clearing, obviously displeased. As he got ready to go, Utley mentioned that Hyde Park's county commissioner would arrive later to explain the progress of some ditch work on Dan Bowles Road. He also asked that one of the deacons present dismiss the group at the end of the meeting. Melvin Stewart and I conducted the rest of the meeting. The next speaker on the agenda was

Terence Dicks, a local African American community organizer who did not live in Hyde Park. Dicks, a tall, gangly man in his late thirties, had been a grassroots organizer for more than eleven years. Dicks described himself as "radical" and was a social justice activist-at-large. He had helped HAPIC with the Masters week protest in 1994 and had continued his involvement since that time. He was now considering running for county commissioner against the incumbent. Dicks made a short speech, commending residents for their patience and determination. He also assured them, "Y'all's story is getting out. You don't know, but if you go around to North Carolina, South Carolina, people know about Hyde Park, and they are taking strength from your struggle." Just as their original bylaws stressed developing connections with the rest of Richmond County, and the success of the Masters week protests was evaluated partly on the basis of the publicity it generated, Dicks's statement underscores the value residents placed on having their plight recognized by the outside world.

After Dicks sat down, I broached the topic of the Brownfield grant. Before I could begin explaining it, David Jackson raised his hand and asked when someone would do something about the rats coming from the junkyard. Another woman asked why the money that Goldberg was making from selling off the scrap metal in his yard could not be used for the neighborhood. Arthur Smith responded to her, agreeing that the neighborhood should have a chance to profit from the junkyard. Smith rose from his front-row seat and addressed the meeting:

> They should hire every young man and young woman who wants to work to clear the site out. We should buy both sides of the property. . . . We should become businesspeople. We should be the ones to control everything happening in the junkyard.

Smith's ideas about neighborhood control of the junkyard generated mixed reviews. Some people clapped for him and nodded their heads, but others sat in silence. One woman said softly, "But we don't want a junkyard." Another added, "They should be paying us." Despite disagreement over Smith's remarks, his ideas about community control of the junkyard affirmed his faith in the potential power of the community members to act together and control their own lives.

Johnnie Mae Brown brought the discussion back to the Brownfields grant and its effects on the community. Leaning forward in her chair, she said, "I thought the Brownfields Initiative was to relocate the community." I explained that the current grant only tested the site and determined cleanup costs. Once the site was tested, HAPIC or the city would need to obtain other funds for cleanup. "How long does it take for them to check for contamination?" asked one resident with a vexed expression. This question inspired collective grumbles about the number of tests done over the years and the lack of results. One man demanded, "Why are they spending the money to be testing out here? They been testing out here for the last ten years; there's enough testing."

David Jackson, who had come to the meeting on his way to band practice, held his hand to his chest and replied, "My daddy used to raise sweet potatoes out here like that, but now you raise something, it rot before you. Something is wrong." Jackson's comments, punctuated by the now-familiar refrain, "something is wrong," initiated a series of stories about people's experiences with contamination and complaints about the everyday effects of chemicals on their health and property. Finally, an elderly man capped the collective grievances, gesturing emphatically with his cane,

> Everybody else got paid but us. Why should we have to pay? The government seem to be the one that's getting all the money, so why can't the government clean this thing up instead of taking it out of our pockets?

Next the meeting attendees turned to Melvin Stewart and me, protesting our inability to answer their questions satisfactorily. One woman asked, "Why did you have a meeting if you can't answer the questions? Bring the people who can answer the questions." This statement prompted a series of loud remonstrations at the fact that the commissioner had failed to appear. Johnnie Mae Brown said, "County commissioners never come to Hyde Park for anything unless they think there's money involved." Richard Johnson stood and admonished the crowd, "You vote these folks in."

The meeting ended according to HAPIC tradition, with the singing of a hymn, but no one seemed very satisfied. The center that night seemed saturated with the frustration and anger that come from long years of waiting, disappointments, and hopelessness. I myself left feeling depressed and helpless. The following day, I received a phone call from Terence Dicks, whose point of view was different. Dicks told me that all

Johnnie Mae Brown speaking at a Brownfield meeting, 2004. Photo by author.

the noise people had made at the meeting was a good thing, "It means they're getting comfortable with you." Dicks's comment prompted me to reconsider the past evening's events.

Indeed, the meeting *had* provided residents with a few things. It was a forum for them to voice opinions and to discuss ideas about issues ranging from economic independence to corporate, civic, and state responsibility—not to mention their critiques of some of their leaders. It was also a place where residents could remember the beautiful gardens their parents had once grown and the hardships they had faced since having to let those gardens die. Expressing these memories and complaints reinforced residents' shared history, especially the ways in which they had come together throughout the years in a collective effort to improve their lives. The memories also reflected the recognition that HAPIC's new problems were, in many ways, reformulations of the traditional civil rights issues they had always battled. Even if the meeting did not produce concrete outcomes or seem to bolster residents' faith in the future, it did provide them with a forum in which to express freely their critiques of government agencies and their shared experiences. And it provided an opportunity for individual voices to be heard. Thus, it embodied the very democratic ideals that formed the cornerstone of their organization. These values, along with their shared experiences, provided a basis for residents' sustained activist identities and consciousness, which they mobilized into action when crises arose. Meetings like the one in August 1999 might seem unproductive at first glance. But if we look at them more closely, we see how in preserving strong community bonds, they produced the stuff of "quiescent" politics, or the underpinnings of a sustained activist consciousness.

Conclusion

Environmental justice scholars have well noted the deep connections that environmental justice activists make between environmentalism and civil rights, and the degree to which those connections make up the frameworks from which they stake their claims.[79] But it is equally important to recognize that within such framings, activists create multiple strategies and identities that are fluid and contingent. Aside from the challenge of fighting environmental injustice, HAPIC activists faced the internal challenge of sustaining their organization through its successes and disappointments, and through changing economic and social circumstances. Moreover, although social movement theorists tend to assume that social movements are unitary, cohesive agents, lived experience tells us that as they move through time, movements ebb and flow, and collective action responds to various social, economic, and historical processes.[80] In other words, people and the movements they create both change with and react to history. Social movement actors develop a variety of strategies (political, economic, and social) so that they can gear their protests to the many corporate and state agencies with which they must interact.

However, the role of the state as a mediator in Hyde Park's struggle was as complex as it has always been in the lives of African Americans. At times, state institutions protected residents from corporate harm, but at other times, they seemed to exacerbate the neighborhood's problems and even to be complicit in corporate deeds. Accordingly, HAPIC activists shifted between appealing to state agencies for help and contesting its biases. In a way, these shifts paralleled their use of science: while recognizing that scientific "facts" are subject to personal biases, activists also recognized that science must be contested on its own terms. Each of their strategies required that HAPIC activists draw on slightly different organizing identities or that they cast themselves in a slightly different light. Thus, their strategies and their identities had always been, and always will be, inseparable and intertwined.[81]

Adaptive strategies do not tell the whole story, however; nor do large-scale protests or marches. Rather, if we are to understand social movements in any significant way, we must look beyond the demonstrations and events that punctuate them and search for the crucial networks of relationships that underpin and nurture collective action before, during, and *in between* events.[82] In the case of Hyde Park, certain fundamental ideals on which residents had always based their activism (i.e., forging greater external connections on local, state, and national levels, honoring local knowledge, and seeking equal participation and equal opportunities) never wavered. Equally important, no matter how their present circumstances changed in economic or social terms, HAPIC activists shared the joys and hardships of having grown up in Hyde Park. Thus, despite the multiple, strategic, and contingent ways in which HAPIC activists presented their identities to outsiders and the internal conflicts that sometimes plagued them, they sustained a strong collective self-portrait, which cast Hyde Park as a powerful and able community of people. It was this self-portrait, and the very act of painting it, that solidified the bonds holding HAPIC together.

Indeed, in the simple act of coming together to voice both the positive and the negative reasons for their organization's existence, HAPIC activists engaged in quiescent politics, nurturing an activist consciousness that in turn nourished an oppositional infrastructure, ready to coalesce into concerted collective actions. For example, in 2004, after they found out that a new recycling facility was being built across from the former Goldbergs site, residents mobilized quickly, appearing in the city council chambers en masse until council members agreed to rescind the facility's permit.[83] Finally, discourse and participatory democracy are most obvious in situations where people have traditionally been disempowered.[84] For the African American activists of HAPIC, who have especially faced (and continue to face) a great number of historical exclusions from the democratic process, democratic organizing meant active participation among all its members. Whether that participation meant strategizing, griping, sharing memories, or even dissenting, HAPIC members made sure that enough time was taken for all voices to be heard. For, as one activist once told me, "That's what democracy is all about."

Six

Staying on Board

I visited Melvin Stewart one evening in his ranch-style house in South Augusta, a few miles from Hyde Park. Melvin, whose boyish face and infectious laugh made him seem closer to thirty than his actual age of fifty-four, had never actually lived in the Park, but he had been HAPIC's treasurer for nearly ten years. As we sat at the kitchen table and discussed Melvin's history and involvement in HAPIC, the dimming sunlight outside the window cast a pinkish hue on miles of trees and fields that stretched into rural southeast Georgia.

That view also led to Melvin's past. Melvin was born in 1949 in Sylvania, Georgia. His father worked in a sawmill, and his mother worked cleaning houses and the all-white high school. But Melvin was not to spend his adulthood in Sylvania. Just after his high school graduation, he was drafted into the Vietnam War. He recalled,

> *Vietnam was an eye opener. . . . You could see it on TV all the time, but it didn't really dawn on me that people were getting killed until I got off the plane and thought, "Oh Lord, this is a real world." Being African American, I started looking around, and I began to see things in Vietnam. After joining my company in the jungle, it appeared that blacks and Hispanics were the majority of 150 soldiers. Normally a tour in Vietnam would last twelve months. I also noticed that a lot of white guys would get "rear jobs" after about six months in the field. A soldier has less than a fifty-fifty chance of getting killed during the one-year tour, if he was infantry. The longer you stay in the field increases the probability of your being killed. I can remember after joining my company, I met several Hispanics and black guys who had been in the field eleven months. They left the field about five to six days before their actual flight home.*

Melvin was surprised by the racism he found in Vietnam, but in some ways it was not as bad as the racism back home. When he returned to Sylvania, it was "like a culture shock. In Vietnam, everyone was depending on everyone else to live, and when I got back here it was like business as usual."

In 1970 in Screven County, Georgia, "business as usual" meant that blacks and whites used different doors in restaurants and stores.

> *It just made you feel, "Why did I go over there? What was I over there for?" And then you started to feel guilty and you started to feel stupid. Like, why did I go over there and risk my life?*

Taking advantage of the GI Bill, Melvin enrolled in Savannah State University. He majored in sociology and then went on to get a master's degree in public administration. In school, he was finally able to make some sense out of his recent history. He explained,

> *When I returned from Vietnam, I enrolled in Savannah State University. College was an excellent place to help me readjust from the war. While I was there, I read a lot of stuff about Vietnam. In fact, I was working on my master's in public administration at Georgia Southern University when I chose to write a paper on the aftermath of the Vietnam War. The research helped me to understand what was really happening in Vietnam. . . . It didn't really make me angry. I just felt relieved.*

Melvin's social science degrees then set him on a life course for social action.

After school, Melvin moved to Augusta and landed a job as a carpenter's helper with the Economic Opportunity Authority (EOA). Shortly thereafter, he took a position as EOA's housing director, where he befriended Mary Lou Utley. A few years later, Melvin moved on to Georgia Legal Services and then Paine College, where he worked as an assistant director of financial aid. At Paine, Melvin ran into Charles Utley. Although Charles eventually left Paine, he and Melvin remained good friends. By that time, Mary Lou had passed away, and Charles had taken her place as HAPIC president. When he found out about his neighborhood's contamination in 1990, Charles told Melvin, who quickly signed on to help.

Charles and Melvin's tag-team leadership style led HAPIC to some unexpected alliances (with, among others, Physicians for Social Responsibility and antinuclear activists in both the United States and Russia)—and to concrete, at-home successes. In his kitchen that night, Melvin reflected on why his partnership with Charles and HAPIC had worked out so well:

> *We really complement each other in a lot of ways. Charles is the hands-on, up-front person. . . . I operate more in the background and mention things to him that I think will be helpful. I think he appreciates that. He seems to trust my opinions, so I just stay out of his way and fit in when and where I can. I enjoy working with him. I enjoy working with the Brownfields stuff. We sit around and talk, and I share things with him he hadn't thought of and vice versa.*

Melvin leaned back and gave me one of his easy grins, "It kind of goes around like that."

A tall man with a smooth, resonant voice, Robert Striggles strode into the Utley Center and immediately went to the back office to say hello to EOA employees Sharon and Diane. He had known Sharon for almost forty years and Diane for at least twenty-five. Although Robert had not lived in Hyde Park since the mid-1980s, he came back often both to visit his sister and nieces who lived on Golden Rod Street and to attend HAPIC meetings. Dressed in khaki pants and a plaid button-down shirt, Robert relaxed into a metal folding chair and clasped his hands behind his head, perhaps considering how many memories the center held. Clearing his throat, he recounted some of them.

"I was born here in Augusta in 1947," he began. "I moved to Hyde Park in 1954. We had outdoor johns then, and everybody had pumps." Those conditions were not unfamiliar to the Striggleses; like most Hyde Park residents, their family originally lived on a farm, near Waynesboro, Georgia. Unlike most Hyde Park residents, however, Robert's father's family had not had their forty acres and a mule stolen or repossessed. "We had a few [family members] that did sharecrop, but the majority of the family owned their own land," he said. Because it was small, the Striggles farm never became very lucrative. To make ends meet, Robert's great-grandfather and grandfather dug wells and preached. Preaching was passed on to Robert's father, who also became a minister. After moving to Hyde Park, he supplemented that work with

a job at Babcock and Wilcox industrial ceramics factory. In addition, on weekday evenings, he ran a barbershop. Hard as his father worked, Robert remembered that the family always had dinner together. "One thing about my dad," he related, "every day he left the barbershop and came home and we ate together." Meanwhile, Robert's mother cooked most of those meals and took care of her children full-time. In addition, she, Mary Utley, and a few other women from the neighborhood began a day care center on Willow Street in the late 1960s. Robert's memories of life within his neighborhood may be positive, but outside of it, he was still living in Jim Crow times. He graduated from the all-black Lucy Laney High School and then in 1966, like many of his peers, was drafted into the service and went to Vietnam.

After surviving the military, Robert returned to Hyde Park and enrolled in Augusta Business College. He also took a job with Lowe's Hardware and worked there for the next twenty-seven years. "I wanted to be the best," Robert remembered. And indeed, within his first fifteen months, he was number ten in the company in retail sales (not to mention one of the store's only African American employees). After several promotions, Robert retired from Lowe's and started his contracting business. Eventually, he and his wife (who worked as a bank teller) saved enough money to expand their house. However, they were unable to buy the lot next door to them, so in 1986 they moved to another neighborhood south of Hyde Park.

Even though he had relocated, Robert carried on his parents' tradition of neighborhood activism and kept his position as an HAPIC advisory board member. He attended as many HAPIC meetings as possible and helped out however he could—sometimes just by supporting an activity such as a computer training seminar (even though Robert himself used computers regularly) and sometimes by participating in or planning a protest. Robert may have earned enough money to move out of Hyde Park, but it would certainly be inaccurate to say that he deserted it. Indeed, like Melvin, Robert represented one of the many nodes in HAPIC's ever-expanding network.

Crossing Murky Waters

On an unseasonably hot Saturday afternoon in early April, I bowed my head along with approximately forty other people in the main room of Hyde Park's Mary Utley Community Center.[1] In his resonant baritone, Reverend Charles Utley led a prayer over lunch. Utley stood with his back to an LCD computer image projector and a wide screen and faced the unusually large crowd of Hyde Park residents who filled the community room that day. Dressed in "Saturday clothes" (sport shirts and khakis, or shorts and T-shirts), they sat at two rows of three long tables, waiting for lunch to begin. We were gathered at the community center for a free workshop entitled "Environmental Justice and Public Participation through Electronic Access." The Urban Environment Institute (UEI), a nonprofit organization based in Washington, D.C., had organized the workshop and intended it to be a first step in developing an environmental justice community technology center in Hyde Park. The UEI's idea was to set up such centers across the country where people could use computers to research local environmental issues: once people obtained the knowledge they needed, they could participate in environmental decision making.

As Hyde Park residents piled paper plates with fried chicken, salad, and cookies, I heard them muttering that the workshop was not turning out the way they had hoped. They told me that the morning had consisted of two presentations by Department of Energy (DOE) staff members explaining how to calculate chemical doses and identify chemical risks. Residents were disappointed that the presentations were overly technical and hard to follow. One DOE staff member later confided, "I'm afraid I bored everyone to death." After lunch, things did not get much better. A UEI staff member explained how to use Geographic Information System (GIS) software to compare neighborhoods' environmental and health conditions. Residents listened politely but shifted in their chairs. Quietly, people started to rise and file out of the room. By 1:30 P.M., only a handful of people remained.

By joining environmentalism and civil rights, the environmental justice movement opens gateways for alliances between African American grassroots activists and white or middle-class (often both) environmental professionals. These alliances present promising new opportunities for grassroots groups like HAPIC to access the funding and political connections of more powerful organizations. The flexibility of the "environment" as an organizing concept facilitated such alliances, and HAPIC activists often used it strategically, as a way to describe their common ground with other groups. Yet, although environmental justice causes often bring together disparate groups, long-lasting and meaningful partnerships are usually difficult to come by.[2] While on an abstract level, HAPIC activists and the professionals with whom they worked shared a vague vision of "environmental justice," I found that the ways in which they defined that term, and the day-to-day steps they took toward achieving it, varied widely and required significant negotiations. For not only did both sets of activists' specific histories shape their present attitudes toward the environment, but they also influenced other critical aspects of social movement organizing such as diversity, computers, technology, and religion.

Importantly, UEI's founder, John Rosenthall, was an African American man who hired mainly African Americans to work with him, in contrast to the mostly white staffs of the professional environmental organizations and agencies with which HAPIC worked. But the "disconnects" and negotiations I am describing *cannot be mapped in any simple way onto race or class differences alone*. On the contrary, the mostly black UEI staff members held the same beliefs about the benefits of computers to environmental justice organizing as white professionals. Thus, although liberal, free-market ideologies present the Internet as a great equalizer, various combinations of social hierarchies and experiences structure people's experiences with computers and with technology in general.[3]

At the same time, in the process of negotiating and working through their differences with professional groups, HAPIC members strategically inserted their ideas about organizing into mainstream discourses and practices. In the end, HAPIC's disconnects with professionals were subtle and significant but not insurmountable. In fact, both sets of activists worked hard to find common ground from which to organize.

Separate and Not Exactly Equal: Environmentalism and Civil Rights in the United States

In 1970, Douglas Moore of the Washington Black United Front called environmentalist Ralph Nader "the biggest damn racist in the U.S." Moore went on to claim that Nader was "more responsible than any man for perverting the war on poverty to the war on pollution."[4] Around that time, Richard Hatcher, the mayor of Gary, Indiana, argued, "[The] nation's concern with the environment has done what George Wallace was unable to do: distract the nation from the human problems of the black and brown American, living in just as much misery as ever."[5] Activists of color have often complained that the enormous popularity of environmentalism as a contemporary issue has diverted political attention away from more typical minority concerns such as housing and unemployment. As a result, civil rights movements and environmentalism emerged in U.S. history as two distinct—and even adversarial—social movements. In the latter half of the twentieth century especially, environmentalism became a highly professionalized movement that pushed an agenda largely geared toward keeping toxic chemicals away from white, middle-class constituencies.

This "second wave" of environmentalism began after World War II and the publication of Rachel Carson's *Silent Spring* in 1962. Environmental agendas shifted from conservation and preservation to critiquing industrial practices and the production of hazardous chemicals. Environmental groups geared toward litigation, lobbying, and technical evaluation proliferated throughout the 1960s and 1970s. During this time, a core sector of environmental groups, known as the "Big Ten," came to define contemporary "mainstream" environmentalism. That core (including groups such as the Sierra Club, the National Wildlife Federation, and the Nature Conservancy)[6] had surpassed others in membership size and access to funding sources. In 1981, representatives from these groups met in Washington, D.C., and decided to combine their forces and divide up their lobbying efforts. The strategy has proved extremely effective: the groups (especially in policymakers' eyes) are generally recognized as representing "the environmental point of view."[7]

Today, a wide range of environmental groups in the United States represent a broad spectrum of interests, goals, and memberships. For my purposes here, however, I define mainstream or professional environmentalists in opposition to localized grassroots environmental justice groups like HAPIC. Whereas mainstream groups have professional staffs, localized grassroots groups tend to be run by volunteers (most of whom hold other full-time jobs). Mainstream groups often have multiple and sophisticated ways of generating funding and gaining access to political clout, whereas localized grassroots groups have limited funding resources and less direct political access. Some of the mainstream environmental groups with which HAPIC has worked include Greenpeace, the Sierra Club, Nuclear Information Resource Services, and the Georgia Air Keepers. In addition, I include in this category the nonprofit foundations that funded HAPIC, such as the Public Welfare Foundation.

Not only did environmental movements historically leave out the concerns of minorities, but as they further institutionalized during the second half of the twentieth century, some of them directly infringed on social justice struggles. For instance, zero population movements (which derived from the Progressive Era and the period just after World War I) resurged in the 1970s and 1990s. These groups responded to contemporary immigrant "tides" by lobbying to reduce U.S. immigration quotas.[8] In addition, when the environmental movement peaked in the late 1960s and early 1970s,[9] African American activists were involved in civil rights struggles, and civil rights and environmental groups often had to compete for foundation and government funding. Because they fought for causes that *appeared* to affect all Americans equally (e.g., leaded gasoline, pesticide use), and due to the already powerful status of their memberships, environmental groups tended to win funding battles.[10]

But the effectiveness of environmental groups had some drastic effects on people of color. For example, legislation that made the siting of new hazardous waste facilities more difficult allowed affluent white neighborhoods to harness stiffer environmental regulations and oppose permits for such facilities in their neighborhoods. "Not in my backyard" (NIMBY) became a rallying cry for many middle-class whites who refused to allow garbage dumps, incinerators, and landfills into their neighborhoods. These facilities, which had to go somewhere, then ended up in poor, powerless communities of color. Or, as sociologist Robert Bullard puts it, public officials and private industries "responded to the NIMBY phenomenon using the place-in-blacks'-backyards (PIBBY) principle."[11] Minority communities have thus become the prime targets for solving "facility siting gridlock."[12]

In addition, people of color tend to hold blue-collar jobs and thus may also see environmentalists as a threat to their livelihoods and, in some cases, their safety. For instance, in the 1960s and 1970s, mainstream environmental groups protested the composition of pesticides, which were highly toxic. In response, pesticide manufacturers changed their products so they would degrade rapidly. However, upon initial use, these pesticides can actually be more toxic than the previous versions. The farmworkers (usually immigrants) who spray the pesticides are thus exposed to higher levels of toxicity than they were before. Further, when it comes to environmental hazards of almost any sort, minorities receive a "double whammy": they are exposed to greater risk and also have less access to health and medical facilities.[13]

In addition to bearing the negative effects of environmental legislation, communities of color have also criticized mainstream environmental groups for their lack of minority representation. A 1988 study, for example, showed that of the Big Ten environmental groups, only six had people of color serving on the boards of directors. In addition, only 16.8 percent of these groups' employees were racial or ethnic minorities. Of those, all but 1.8 percent worked in administrative or "blue-collar" roles.[14] After one hundred grassroots environmental justice leaders signed a letter to the Big Ten, accusing them of racism in their hiring and policy practices, many environmental groups did work to diversify their staffs. But, while more radical environmental groups, such as Greenpeace, started environmental justice initiatives in the early 1990s, it was not until the late 1990s that the Sierra Club's new executive director announced, "[The environment] is not just about Yosemite and the beauty of the wilderness. It is about cities—the air we breathe and the water we drink."[15]

By the end of the 1990s, an increasing number of mainstream environmental leaders were eager to change their reputations as exclusive, largely white, middle-class organizations by setting environmental justice priorities. Mainstream groups needed the credibility (and moral clout) that came from working with grassroots, minority groups. In turn, grassroots groups began to count on the financial and in-kind support and access to governmental agencies that connections with mainstream organizations gave them. Today, many alliances between mainstream environmental and grassroots environmental justice groups have transformed into symbiotic relationships. For grassroots activists, however, the history of being excluded from the environmental movement still generates a certain amount of mistrust and resistance to establishing those relations. Grassroots activists especially worry about retaining a sense of control over a particular issue or organization. Finally, the ways in which African Americans define the environment may differ significantly from the ways in which it is defined in mainstream American discourse.

But important categories and discourses like race or the environment can be used strategically. The next section illustrates how HAPIC activists sometimes altered their racialized discourses about the environment in order to build and maintain partnerships with powerful white organizations.

Rainbows and "Salad Bowls": Making Environmentalism Multiracial

On a bright October day in 1998, the Georgia Air Keepers Campaign held a press conference in front of the Augusta–Richmond County office building. The Air Keepers, a coalition of Georgia nonprofits with environmental interests, intended to publicize the release of a new and startling report on the dismal condition of Georgia's air quality. Speakers included a young white woman representing the Georgia Public Interest Research Group, a middle-aged white male physician from the local chapter of Physicians for Social Responsibility, and Charles Utley. Press conference organizers saved Utley's speech for last. Stepping up to the microphone, he took a deep breath and began to speak in the best way he knew—cultivated through years of seminary training. In full-scale southern preacher mode and a shivering voice, he delivered a sermon on making sure that the environment was safe for all humans. Ending his speech, Utley cautioned his audience to remember that the environment was "not about black or white, but all colors."

Utley *did* want the environment to be safe for all human beings. However, his emphasis on the multiracial aspects of the environment (how it affected "all colors") contrasted with the racialized language that he often used among other HAPIC members. Utley was not unique among African American activists in this. For example, one activist from Savannah told me, "Really, the movement is about environmental *in*justice and discrimination. We have to bring that out more." But, she added, "people like to use [environmental justice] because it's polite." To appeal to press conference audiences and mainstream groups like the

Georgia Air Keepers, HAPIC leaders thus cast the environment as an ambiguous entity that united all races, strategically shifting their discourse and de-emphasizing the racial bases for their activism.

There were good reasons for this. By the late 1990s, not all environmental groups had effectively diversified their memberships, staffs, or agendas. Many mainstream groups still hesitated to highlight race as a specific issue of concern. When the Sierra Club Legal Defense Fund changed its name to Earth Justice Legal Defense Fund in 1997, for instance, its publicity campaign featured ads with a photo of the group's celebrity spokesperson, Mel Gibson. Alongside Gibson's photo ran a long caption describing a recent legal victory for a small community in Louisiana. The ad's copy explained that a "giant multinational corporation chose [the Louisiana] neighborhood for a dangerous nuclear materials and waste facility, presumably because the residents were poor and wouldn't fight back." Although this neighborhood was also primarily African American, the ad describes plaintiffs as "poor," not poor and black. The ad thus appealed to the sympathies of a wider audience by emphasizing environmental classism, a less loaded topic than environmental racism.[16]

Similarly, HAPIC leaders were savvy enough to know that retaining connections with mainstream organizations meant softening their racial messages. Accordingly, they developed strategies for working with mixed-race groups. For instance, Utley stressed multiracial language in his public speeches, but at HAPIC meetings and among African American community leaders he insisted that Hyde Park's environmental problems had "95 percent to do with race." He and other HAPIC leaders also often referred to their situation as "environmental genocide" or "environmental apartheid" among African American groups. When I questioned Utley about his mixed environmental messages, he explained,

> The environment is going to affect everyone. [Corporations and governmental agencies] try to just get blacks and they think this contamination won't affect them—that it will stop where the black neighborhood ends, but it gets everywhere.

Utley also admitted that in his public speeches, he sometimes truncated this message because "you have to get a feel for your audience." In other words, in mixed-race contexts, Utley chose to emphasize only the first part of his message—that "the environment is going to affect everyone." Thus, he had developed a strategy for appealing to white audiences whereby if he could convince them that they were equally threatened by toxins, they might find it in their own self-interest to help Hyde Park residents. In addition, by casting the environment as color-blind and by de-emphasizing their racial identities as well as racism itself, HAPIC leaders appealed to the broader concerns of professional environmental groups that wanted to appear diverse in membership.

Indeed, over the past thirty or forty years, a number of intersecting social movements and trends have made ethnic and racial diversity politically popular.[17] HAPIC leaders, who had established numerous national, state, and local networks with politicians, governmental entities, and other social movement organizations, were well aware of the value of multiculturalism's political currency. By emphasizing diversity, HAPIC cannily increased its chance to win political favor.[18] HAPIC leaders, then, developed a complex environmental narrative that at times stressed their racial victimization and at other times emphasized the environment's potential to threaten all races alike. Thus, in addition to including all the social resources to which Hyde Park residents needed access, the environment could be extended to encompass coalitions with powerful mainstream groups.

But narrating unity with mainstream activists was not the same as uniting with them on a day-to-day basis. While HAPIC leaders worked hard to describe a shared agenda with professional environmentalists, the path to pursuing that agenda was not always smooth. The next section focuses on how HAPIC activists and professionals differently understood the role of computers in social movement organizing.

Stepping into the "Digital Divide"

In the late summer of 1999, I sat around a long folding table with six African American activists from Augusta. This time, I was in the Gilbert Manor public housing community room as part of a group that was working on improving Augusta's transportation system. Like many people around the globe, the other activists and I were chatting about Y2K (or whether or not computers would be able to integrate a new millennium into their operations). Mrs. M., a resident of Gilbert Manor in her sixties, with graying hair that often escaped from her hairpins, exclaimed, "Girl, you better get you some fatback and some beans come January 1st, 2000. 'Cause Y2K is coming."

Mrs. M. had never used a personal computer, and like many activists with whom I worked in Augusta, she was reluctant to start. Contrary to popular assumptions, I found that such resistance crossed occupational and income levels, and it was independent of computer access. Importantly, I do not suggest in any way that reluctance to use computers indicates that people were incapable of learning about them—on the contrary, some of the people referred to in this chapter who were unfamiliar with computers in 1999 are now quite up to speed with them. Rather, I point out some of the practical and more fundamental reasons for an *initial reluctance* to use such technology on a day-to-day basis.[19]

At the end of the twentieth century, connecting minority communities to computers, or bridging the "digital divide," has become a national priority. In 1999, for example, the U.S. government initiated a number of federal efforts to address the problem. Also in 1999, AT&T, AOL, GTE, Compaq, and Microsoft all promised donations ranging from $500,000 to $1.2 million to nonprofits that would bring computer access and training to economically disadvantaged Americans.[20] For professional environmentalists, bridging the digital divide was crucial to the "capacity building" or "empowerment" of environmental justice groups. For example, in the view of John Rosenthall of UEI, the future success of environmental justice struggles depended on computer technology. In "Environmental Justice and Public Participation through Electronic Access," a booklet written for the UEI workshop, he wrote,

> Access to, and the ability to use computers and the information on the Internet will, to a great degree, define those who will have access to economic opportunities and power in shaping public policies, and those who will not.

In the same booklet, a DOE official echoed Rosenthall's view on the importance of computers to grassroots organizing:

> During 1998, we worked with Howard University's Urban Environment Institute to donate computers to communities around the country. The purpose of this effort is to facilitate community access to environmental information and to increase their ability to participate in envi-

> ronmental decision-making. . . . Hopefully, the community technology center at the Mary Utley Community Center will be a model for environmental justice communities countrywide. . . . It is our intent for these centers to build capacity for public participation.

Thus, for professionals, computers and the Internet presented a thoroughfare for accessing power in the United States.

HAPIC leaders were excited about the prospects that computers brought to organizing, at least *theoretically*. Utley's contribution to the workshop booklet stressed the number of opportunities that computers created for HAPIC as an organization:

> We can do a lot with these computers. We can conduct basic computer classes for adults and children. We can conduct environmental research to keep up with and participate in the environmental justice movement. We can learn about what is going on at the Savannah River Site,[21] and give them our comments. We can seek funding sources. We can communicate on a regular basis with our elected officials in City Hall, the Capitol, and in Washington. We can network with environmental justice activists around the country, to know what works and who works for us. All of these possibilities will contribute to a much better life for our community.

Utley's high hopes for what computers could do for HAPIC focused on the ways in which they could develop connections to governmental entities—one of HAPIC's organizing goals since its inception in 1968.

However, when Utley got funding for two brand-new computers and training classes at the Utley Center in 1999, the program was not instituted for several years. In the meantime, he continued to conduct HAPIC business as usual, holding community meetings, cleanups, recreation programs for neighborhood children, and so forth. There are several reasons for Utley's hesitance to initiate the program. First, he worried that new computers might get stolen from the center. Second, with his full-time job as a guidance counselor and his part-time work as a minister, Utley was short on time. Third, computers were not essential to HAPIC's day-to-day activities. Fourth, a vague idea of shaping public policy in the future did not solve HAPIC activists' immediate problems of toxic poisoning. All these factors suggest that, ultimately, technology and training alone will not suffice to bridge the digital divide.

In fact, in 1998, learning about computers, and using them for communication, was becoming increasingly pressing for HAPIC leaders, who wanted to build relationships with professional organizations. Professionals relied on computers for such tasks and had strong convictions that HAPIC activists needed to do the same. For example, one of HAPIC's funders required that grantees submit expense reports and budgets using spreadsheets. Utley had entered the information into a word-processing program, but transferring it to a spreadsheet program would require a lot of time, especially given his limited facility with computer software. In addition, although Utley kept meticulous receipts for each HAPIC check he wrote, he had not organized the receipts into the same categories that the funder was requesting. After several weeks of budget revisions and reworkings, the grants manager suggested that HAPIC include money management software in its next grant. HAPIC leaders agreed to the idea but never bought the software and continued to organize their finances as they always had.

In a second example, activist Arthur Smith took a position on the board of Waste Not Georgia, a statewide recycling organization. Because it was a statewide program, Waste Not conducted almost all its business via e-mail. At first Smith promised to learn how to use e-mail (he had several places where he had access to free training and computers) and to stay current with board business electronically. But after several months he said he was too busy and asked board members to telephone and airmail him all the information he needed. Similarly, in 1999 and 2000, Charles Utley, who had a computer at home, picked up e-mail messages from John Rosenthall, his daughter, and some other mainstream environmentalist groups with which he worked. However, Utley usually responded to e-mails with telephone calls. While HAPIC activists generally used faxes to transmit printed materials such as flyers announcing meetings and conferences, their preferred mode of communication was almost always the telephone or face-to-face contact. As the years since my fieldwork have worn on, Utley has come to use e-mail more, but the examples here show that having easy access to a personal computer or basic knowledge about how to use it made little difference in whether grassroots activists *embraced* it as a medium of communication.

In 1999 I, too, jumped on the digital divide bandwagon and organized a free computer training class for Hyde Park adults. Initially, many residents expressed interest, and I arranged class times according

Computer class at the Mary Utley Community Center, 1999. Photo by author.

to when people said they were most likely to be available. But, on the first day I had only two students. One of these was Daisy Mae Burden, a stout grandmother of four in her middle to late sixties who leaned heavily on a "stick." She explained why she wanted to take the class, "Even the supermarket's using [computers] now. I better get with it or I'm going to be left behind." Burden also hoped one day to communicate with her grandchildren via the computer. My other students, almost all of whom were women in their fifties and sixties, also said that they wanted to use computers to communicate with their children and to "be part of things."

Yet class attendance remained scant, and over a five-month period, only a total of five residents ever came. Eventually, I began to question Hyde Park residents and activists about their reluctance to learn about computers. In answer, they pointed to a variety of factors. Some said that older folks tend to be afraid that they will mistakenly erase files. Others pointed to the fact that many neighborhood residents could not type very well, or their spelling was poor. Communicating via e-mail, then, could be laborious and expose poor composition skills.[22]

For a number of people, computers also initially threaten the deeply held values about social movement organizing that formed the basis for their social action. In the case of HAPIC, computers, and technology more generally, represented some of the very exclusions against which they had historically fought.

Technology as a Language Virus

On my initial visit to Hyde Park in January 1998, I met several HAPIC board members for lunch at S&S, a traditional southern-style cafeteria. While we dug into our catfish, okra, and cornbread, my companions related a tale I was to hear several times during my fieldwork. It concerned the meeting at Jenkins Elementary School at which EPA and ATSDR representatives announced the results of their $1.2 million study. With a collective chuckle, my lunch mates described how the meeting culminated in one resident's throwing a chair onto the Jenkins stage.

For the first thirty minutes of the meeting, EPA officials had discussed concentrations, key contaminants, and sample data, but residents still had no idea whether or to what degree their environment was contaminated. Charles Utley finally broke in and said,

> You need to know one thing. All of your data says, primarily, something we can't even comprehend. Secondly, what is happening to the people? Why is this going on? Why are there respiratory problems like we have? Why are people dying of circulatory [diseases]? That is what they want to know. They don't want to know how much arsenic per million, per billion, per trillion.[23]

After waiting more than a year for the study results, those at the meeting were more than a little perturbed when officials communicated those results in incomprehensible language. It is not surprising that tensions in the room finally exploded into the chair-throwing incident.

At some point, environmental issues, particularly those dealing with contamination, become scientific, and HAPIC activists found that discussions about their situation often concerned specific chemicals, parts per million, hydraulic flows, and point source pollution. Although many HAPIC activists had developed a sophisticated scientific vocabulary by the time I began fieldwork in 1998, what I refer to as "techno-speak," or a heavy reliance on technical jargon, continued to be a major source of residents' feelings of alienation from the professionals who entered their community. HAPIC activists are not alone in this regard—many environmental justice activists similarly report being irritated by the overuse of technical talk by various governmental officials.[24]

HAPIC activist Robert Striggles explained some of the reasons for the extreme frustration that he and other community members felt at the meeting:

> I was over at the school and [laughs] they wasted, I believe two million dollars—what the study cost them—which should have been used in the area to relocate the people. When you started coming in dealing with numbers they was talking about, so many millions, . . . The peoples couldn't relate to it. All they wanted to know at the meeting was there's contamination or there's no contamination. . . . Even if you was up on all [the lingo] you still couldn't understand what they was talking about. . . . That's what I think was the biggest problem because it was so far over everyone's head unless that was your background, you didn't know exactly what they was talking about.

I asked Striggles why he thought EPA officials had insisted on using language that went "over everyone's head." He replied,

> I think [the EPA officials] looked down at the area. That's exactly what it was, in my opinion. And the reason was, if you come into this area and you're giving them statistics that they don't understand, that they had never been dealing with, then what are you doing? You're looking down on them.

When I further questioned him about why the EPA looked down on Hyde Park, Striggles responded simply, "This is a minority area." Due to their history of exclusion from state-level decision-making processes, Hyde Park residents already harbored substantial mistrust of government officials. For them, speaking in incomprehensible language only confirmed and revealed the officials' latent racism. Anthropologists have described such magnifications of mistrust as "risk perception shadows" or "a predisposition to distrust projects involving potential adverse health or social impacts and to doubt agency or company statements regarding the potential dangers associated with these projects."[25]

Given Hyde Park residents' historical experiences as poor black southerners, such shadows certainly loomed over them and were only darkened during the risk communication process.

As a result of the notorious community meeting, relations with the EPA remained wary and somewhat adversarial. Even after President Clinton's executive order on environmental justice, technical talk continued to create problems between governmental agencies and Hyde Park residents. For instance, in 1998 DOE officials working at the Savannah River Site (SRS), a nuclear weapons facility approximately twenty miles away from Hyde Park, visited the neighborhood. At the time, SRS was lobbying for a grant to process plutonium, and part of that process required it to solicit community approval of its intended operations. Arthur Smith described SRS officials' failure to beguile the community. He said, "SRS came out to educate the people of Hyde Park about the plutonium process. But they put it over the people's heads." Smith later explained that SRS officials "were out of touch" with the community. By not trying to "speak their language," these officials confused people and prevented them from gaining information. Although Smith did not directly name this process as a racist one, his interpretation of it implied a willful obscuring of information and paralleled Robert Striggles's ideas about governmental officials "looking down on the community."

The problem of technical talk was not limited to government officials. Many of the professional activists who came to work with HAPIC also spoke in an unfamiliar language that created barriers between themselves and HAPIC members. Techno-speak, then, emphasized the race and class differences between professional and grassroots activists—differences that had led to their disparities in education and job opportunities. Because technical language also excluded them from accessing important information about their health and safety, HAPIC activists folded its use into their experiences of environmental racism. Thus, for them, techno-speak was a mode of communication that both reflected and translated into racism.

Like the idea of the environment, in certain contexts, technical language and computers connoted discrimination—in this case in education, particular types of employment, and other social opportunities. Motivating HAPIC activists to use computers to achieve environmental justice and to increase their power, then, was not merely a matter of giving them computer access and training. In Hyde Park, technology did not represent the obvious road to empowerment that professionals assumed it did because it was circumscribed by complex social inequalities. In addition, computer use did not necessarily jibe with HAPIC members' traditional mechanisms for establishing and seeking inclusion, participation, and power. Perhaps the most important of those traditional mechanisms was religion.

Jesus' People for Environmental Justice

As Hyde Park residents filed into folding chairs for a late October 1998 HAPIC community meeting, Arthur Smith lifted his raspy, tuneful baritone and began to sing the hymn "Nobody but You, Lord." Immediately, other attendees picked up the tune and sang along until it filled the room and was on everyone's lips—that is, everyone except two professors from Augusta State University, attorney Bill McCracken, two Project South staff members, and me, the only white people in attendance. Later, at the end of the hour-and-a-half meeting, Reverend Utley said a closing prayer as everyone bowed heads. We then joined hands and sang (or hummed) HAPIC's theme song, "Reach Out and Touch Somebody's Hand."

Every grassroots-led meeting that I attended during fourteen months of fieldwork began and ended with a prayer, and if there was food at a meeting, someone blessed it. Almost all HAPIC flyers contain some reference to God or Jesus, and most public speeches included a mention of God. Christianity, specifically black Baptist faith, was an integral part of grassroots environmental justice organizing not just in Augusta but also across much of the South. This section's heading, for example, borrows the name of a long-standing and well-known group in Alabama. I found that a number of southern environmental justice organizations similarly grew out of church groups. Even groups that were founded on a secular basis, such as HAPIC, were built on strong religious roots that could not be separated from environmental activism.[26] For instance, one week in the spring of 1999, Arthur Smith decided to spray-paint the street in front of his house with environmental justice and religious slogans such as "Hyde Park Aragon Park Improvement Committee," "Must Have Environmental Justice," and "Lord Have Mercy." For Smith and for almost all the African American activists I knew in Augusta, religion and social movement organizing were inextricably intertwined.[27]

This way of organizing was strikingly different from the secularism

In front of Arthur Smith's house, 1999. Photo by Maryl Levine.

of professional environmentalists. On the surface, professionals easily accommodated HAPIC activists' religious practices. When HAPIC activists bowed their heads over their food, professionals said "amen" along with them. When HAPIC members clasped their hands at the end of a meeting, closed their eyes, and swayed to a hymn, professionals swayed along in respectful silence, showing careful consideration for the cultural differences of grassroots activists. However, I found that overcoming such differences was not simply a matter of sharing prayers and hymns. Although hierarchical to a certain extent, black Baptist practices emphasize group participation, face-to-face interactions, and the power of the spoken word. Their privileging of these traditional modes of communication explains some of the reasons that HAPIC activists were reluctant to adopt e-mail into their daily practices. Religion was not just embedded in HAPIC activists' experiences as African Americans; it also provided a strong platform from which to organize social action. Moreover, for HAPIC activists, this mode of organizing had met with substantial success.

The close intertwining of religion and activism has historical antecedents. Historian Eugene Genovese, for example, writes that religion prevented the complete dehumanization of slaves by presenting an alternative vision of their worth in God's eyes. After the Civil War, religious spaces continued to function as spaces of inclusion and self-determination.[28] Even during the Depression, historian Robin Kelley argues, religion provided the kind of courage that feeds activism: "A belief that God is by one's side has a significant effect on one's willingness to stand up to a foreman or participate in a strike."[29]

The church itself has also played a controversial role in the history of black activism. Some argue that even though whites kept an eye on southern black churches, as long as preachers were not too radical in their sermons, African Americans could direct their religious institutions in relative freedom from white control. They could not vote in political elections, but African Americans could select their own pastors, deacons, and deaconesses. Moreover, some churches and ministers were instrumental in plotting slave revolts and operating the Underground Railroad. After the 1940s, as African Americans migrated from rural areas to urban centers, urban black churches increased dramatically in number and membership. As a result, they became stronger organizationally and more independent financially, eventually becoming the fertile soil from which civil rights struggles grew.[30] Yet other scholars argue that many churches became quite conservative, especially during the Depression, when black ministers began to separate themselves from politics. Adolph Reed states, "The ministerial practice of 'easing community tensions' has always meant the accommodation of black life to the existing forms of domination."[31]

Environmental justice activists, too, debated the church's relationship to activism. For instance, some activists (often, from larger southern cities) cautioned that churches are susceptible to manipulation. One woman said,

> My only concern is that churches are not used to somehow weaken the movement. . . . We would like to see the church involved if the church is going to be accountable to the people. And too often the church is that one minister who may have economic concerns. Too often we have seen churches break the back of struggles because they're manipulated by the power structure.

Others believed that educating pastors was essential to garnering widespread support for environmental justice. For instance, Reverend Bobby

Truitt often traveled to ministerial meetings and conventions and tried to convince his colleagues to get involved in environmental issues. But he explained that they often did not connect environmentalism with social justice:

> You see when I started talking "environment," there's a number who cut off, you know, they never get to the justice part. So if we can get that, then we can bring in civil rights and environmental justice. But we rarely get there.

Converting pastors into environmental justice activists might have been difficult work, but activists becoming pastors was common.

Indeed, I found that many black activists in Augusta (and elsewhere in the South) were community organizers first and later became ordained as ministers. Reverends Utley, Oliver, Truitt, Lyde, and Cutter all "heard the call" after many years of taking community leadership roles. Terence Dicks, an African American community organizer, explained, "People are changed through the experience of being involved in a highly charged emotional struggle. It transforms you spiritually." Dicks also admitted that, despite having been an activist for at least ten years, once he became Reverend Dicks, he attained a certain amount of legitimacy in the eyes of the community.

These examples diverge from civil rights era organizing patterns, in which leaders were often clergymen *before* they became activists.[32] In his essay "On the Wings of Atlanta" (1903), sociologist W. E. B. Du Bois predicted that the old "leaders of Negro opinion," the black preacher and the black teacher, would be replaced by other kinds of professionals. In some ways, Du Bois's prophecy has materialized. Reforms gained during the civil rights era opened up possibilities for new spaces in which African Americans could be politically active. For example, neighborhood associations and community boards began to offer activists channels of political participation.[33] Thus, in the post–civil rights era, black activism is no longer necessarily headquartered in specific churches but has spread throughout civic organizations.

At the same time, religion continues to play a major role in African American life in the South, especially in communities, like Hyde Park, that have a large number of low-income residents.[34] It is no surprise, then, that many southern African Americans especially bring their religious practices into the civic spaces where they now center their activism. In addition, although most people in Hyde Park were black Baptists, not all southern environmental justice activists practice Christianity. Some favor African-inspired spiritual practices, and a greater number combine Baptist faith and African traditions such as pouring libations to ancestors at church services. What is clear is that all the African American environmental justice activists with whom I worked infused their activism with some kind of spirituality.

This religiosity illustrates one of my earlier points that certain cultural institutions (such as religion), which may not seem overtly political, often foster and sustain grassroots political action. Thus, whether or not environmental justice was closely tied to the church itself, religious beliefs and practices permeated the movement—and supported and bolstered it. Moreover, in many cases (e.g., activists-turned-pastors) traditional religious values underpinned concepts of legitimacy and authority. In turn, the idea that true authority lay with religious leaders, and with a higher power, fed the ways that activists understood categories like empowerment and participation.

A Pox on MOX!

Charles Utley banged his gold-plated gavel on a lectern, adjusted the judge's robes he wore over his dress shirt and tie, and called to order the approximately forty men and women seated before him. It was late February 1999, and we were in a lecture hall at Paine College. "Judge" Utley was presiding over a mock trial on the effects of developing a mixed oxide (MOX) fuel program at SRS. The trial was part of a national program sponsored by the Nuclear Information and Resource Service (NIRS) publicizing the negative impacts of a proposition to recycle plutonium-based weapons into nuclear fuel at SRS. Augusta's proximity to the facility made it one of the NIRS staff members' most important stops. For that night, they had decided to disseminate their information in the form of a mock trial. Other trial "witnesses" included representatives from the Peace Resource Center in Columbia, South Carolina; the Blue Ridge Environmental Defense League in Charlotte, North Carolina; and Save Texas Agriculture and Resources in Amarillo, Texas. All were there to testify to the dangers of the MOX program.

Events at the mock trial tie together many of the issues presented in this chapter. To begin, HAPIC and NIRS were an unlikely pairing.

Unlike environmental justice groups, antinuclear groups tend to be single-issue oriented, middle-class and professional white associations.[35] Indeed, the anti-MOX activists were all white professionals, with the exception of one South Asian scientist and NIRS's South Asian American administrative assistant. Not only was nuclear waste an unusual topic for a group like HAPIC, but also the antinuclear activists had had little exposure to environmental justice issues. For instance, one NIRS activist told me privately,

> I realized the other day that environmentalism and environmental justice are not really separate issues. We're working with environmental justice groups because it's all the same issue.

An alliance with HAPIC, therefore, constituted relatively uncharted territory for anti-MOX activists. At the mock trial, many of the "disconnects" noted in this chapter stood out in high relief.

Of the approximately forty audience members on that February evening, about half were African American. Except for a few Paine College students, almost all had come from Hyde Park. For several weeks, Charles Utley had been posting flyers and making phone calls urging HAPIC members to attend and to "find out what was in their backyard," stressing that SRS was located only twenty miles from Hyde Park. HAPIC's involvement with SRS had been growing steadily over the past few years. As mentioned earlier in this chapter, the year before, SRS officials had visited Hyde Park to solicit community input on its plutonium-processing grant (the same process now being debated at the mock trial). At that time, SRS's interest in Hyde Park stemmed from its need to adhere to new federally imposed environmental justice guidelines. The guidelines, which were derived from President Clinton's executive order, required federal agencies such as the DOE (which owned SRS) to include nearby minority communities in their decision making. Because HAPIC was the only visible minority environmental group in Augusta (in fact, it was one of only two environmental groups in Augusta, the other being the Sierra Club), SRS officials sought its input. Similarly, in seeking local grassroots support for their campaign against MOX, antinuclear groups found HAPIC.

For these groups, the Savannah River Site had long been a source of concern. In 1954, SRS first fired up its five nuclear reactors and became one of the country's foremost producers of nuclear weapons.[36] At the peak of its operations, SRS employed twenty-five thousand Georgians and South Carolinians. Some of these were Hyde Park residents like Arthur Smith, who worked on SRS's security force from 1981 to 1988. In 1989, post–cold war cutbacks forced the site to close the last of its reactors, causing the loss of eleven thousand jobs over a ten-year period. Between 1991 and 1997, Richmond County lost 54 percent of its SRS workforce.[37] As the 1990s wore on, anxieties about winning the cold war transformed into concerns about nuclear disarmament. The disposal of highly explosive, and highly dangerous, weapons-grade plutonium was a particularly pressing matter. In the late 1990s, the DOE announced that it planned to dispose of the nation's plutonium using a dual conversion process. The first process would seal and secure plutonium into hockey puck–sized cylinders and then store them underground, encased in radioactive-insulated material (to discourage theft). The second process would reuse plutonium as nuclear fuel for power plants.[38] Soon after the DOE's announcement, the country's large nuclear weapons plants began competing for the federal project, hoping to regain some of their operations. In 1998, the DOE named SRS as the preferred site for converting the plutonium. Antinuclear groups opposed the generation of nuclear fuel, and the dual process plan raised a great hue and cry among environmentalists across the country, yet the absence of environmental action groups in Augusta combined with SRS's history as a major local employer made Augustans tough customers for MOX opposition. With HAPIC as their main contact, anti-MOX activists hoped to win the support of both Augusta's white liberal and minority communities.

For HAPIC leaders, working with anti-MOX activists enabled them to extend their networks further into the realm of mainstream professional organizers. Anti-MOX groups were well established, well educated, and well funded, and association with them publicized HAPIC's plight. However, HAPIC leaders had to convince their constituents that MOX was an issue of immediate concern. At the trial, Utley explained that plutonium would be shipped to SRS by rail, and since rail lines always seemed to border minority neighborhoods, they would be the most threatened by leaks and spills. He said,

> Let me tell you why Hyde Park and Aragon Park are involved in this. . . . If we're looking at SRS or Pentex, whatever it may be, it is our obligation, it is our dream, to make sure that another Hyde Park [isn't]

> created. . . . And I stand here tonight to say that if there are any problems that [are] going to be coming through our neighborhoods by rail, we want to know about it. And that's the reason that Hyde Park is holding this meeting. It's an information meeting.

Although the anti-MOX activists had already confirmed that plutonium would most likely be shipped by air, Utley continued to postulate that it might come through on rail. In so doing, he emphasized its potential danger to Hyde Park residents and its likelihood to affect "our" (i.e., minority) neighborhoods disproportionately.

Anti-MOX activists shared Utley's concern about presenting the MOX issue to Hyde Park residents in such a way that it matched their interests and experiences. Before their arrival in Augusta for the mock trial, the activists asked HAPIC leaders to schedule press interviews. Utley was able to confirm two: one with the *Metro Courier,* a local African American newspaper, and one with the *Augusta Chronicle.* At the *Courier* interview, anti-MOX activists confessed their anxiety over the upcoming event—so far, all their audiences had been white, and they expected the one in Augusta to be their first mixed-race audience. The *Courier*'s African American editor in chief advised them to focus on how MOX would affect the black community. She suggested that they address the issue of jobs and how their creation would not outweigh the negative effects of MOX. Here, the editor echoes the sentiments of Utley and other HAPIC leaders in recognizing that jobs are not a tradeoff for clean environments. Finally, the editor urged the anti-MOX activists to avoid using too much technical jargon.

Utley also worried about the techno-speak issue. In his introductory statement, he directed his comments to HAPIC members, remarking,

> We want [this hearing] to be an information piece. And you know, for a lot of people to get information, they need to be in a relaxed mind. So I encourage you to relax. When I say that it is information project, you know in order to do anything you have to have fun at doing it and you have to want to do it and I encourage you to want to—let's want to learn this information. . . . So I want you to be able to cross-examine these experts because it is the information that they have that you need.

Yet despite the *Courier* editor's caution and Utley's advice to relax, as the trial unfolded, "witnesses" were unable to stay away from scientific jargon; in turn, audience members shifted in their seats and sighed audibly. At one point, activist Terence Dicks stood and said,

> Immobilization, vitrification. It comes across to ordinary people as code words for power and for money. I hope that this community won't be put off by certain people's use of vocabulary and jargon and expertise, and I'm glad the expertise is here. But there are real concerns in this room, and I hope nobody will be put off as far as trying to get to what these issues are actually about and that they will get to more meetings and try to find out what's going on here.

Both Dicks and Utley stress the need for community members to learn about environmental situations—even if those lessons came wrapped in highly technical words. Once members found out "what's going on," they could insert their own opinions and voices into SRS's activities and protect themselves from potential harm.

Throughout the trial, in addition to asking for members' patience, HAPIC leaders continually related the issues to Hyde Park's immediate and traditional concerns. For example, Terence Dicks stood up again and attempted to make local sense of the anti-MOX presentations in the following exchange:

> TD [referring to more jargon-heavy explanations of the dangers of MOX]: I don't know about any of those things. But I know that Chernobyl happened and people are very sorry about that . . .
> AUDIENCE MEMBER: Three Mile Island . . .
> TD: . . . Three Mile Island. And the Tuskegee experiment, which the president of the United States got up and said how sorry he was about that.[39] How sorry people are. People are sorry afterward. That's why hearing these ideas is very important, and it's very important to the democratic process.

Dicks's bringing up Tuskegee signaled a shared racial history and reminded audience members of how government can prey on uninformed African Americans. His statement also emphasized that for African Americans, the consequence of remaining uninformed was exclusion from "the democratic process."

For HAPIC activists to engage the issues presented at the trial, they needed to incorporate them into already established priorities such as racism and the health of their families. Toward the end of the trial, Hyde Park resident Richard Johnson stood and, with his powerful frame and booming voice, commanded the attention of his neighbors:

> I been around sixty-three years, so what I'm saying, you can't escape from this stuff. You got people hold these jobs; they're only concerned about the money. They're not concerned about the health care of children. That's what we got to start protecting—the future of the grandchildren. . . . The Bible said that we are strong ourselves and we got to act to that end. We got to be active everywhere.

Johnson's comments were met with rousing applause. By mentioning children, the Bible, and corporate villains, he framed the MOX issue as directly related to HAPIC members' concerns. HAPIC leaders agreed with the professional activists that access to information was linked to empowerment. At the same time, HAPIC leaders also ensured that such information flowed through a culturally understandable idiom. In so doing, they inserted issues that were relevant to them into more mainstream environmental narratives. For HAPIC activists, including themselves in the discourses of mainstream activists was as important as including themselves in governmental decision making.

During the trial, HAPIC members also struggled to insert their ideas about fairness into the trial's proceedings, even when it meant disagreeing with anti-MOX activists. For instance, one SRS executive—a man whom I will call "H."—quickly became angry that the trial's only witnesses were anti-MOX activists. He stood and accused Utley of running "a kangaroo court" and showing just one side of the issue. The anti-MOX activists tried to silence him. One witness said, "Judge, you don't have to let this line of questioning go on." Interrupting her, an HAPIC member mumbled, "I want to hear it." A chorus of nods from his fellow members confirmed this desire. "Judge" Utley responded diplomatically,

> Let me just say that we want everyone to be heard. There may be some differences, but this is an opportunity for everyone to hear. And if we have our own agendas, then we need to put our own agendas aside. . . . If Mr. H., Dr. H.,[40] has some information, I want to certainly hear it, but I don't want it in an adversarial way.

A short while later, H. tried again to raise his objections to witnesses' testimony, and the witnesses again tried to silence him. This time, Richard Johnson stood up and addressed the auditorium's small stage. To the trial "witnesses" he said,

> You took up fifteen minutes of time. And I'm aware of this—I know that you all got a point that you want to get over, but we came here, too, because we have a point that we want to get over. So I want to hear what the doctor got to say also. . . . So, we're here tonight to try to find out what's going on about Savannah River plant. I do want the doctor to finish his conversation 'cause he had a very strong conversation to me. I would like to hear the rest of it, please.

Here Johnson stresses the importance of giving H. equal time to speak and present his side of the issues, emphasizing certain democratic ideals that HAPIC members held dear and that I will discuss presently.

Thereafter, each time an HAPIC member asked a question or made a comment (and most made several), he mentioned either "freedom of speech" or "letting the man speak his piece."[41] Not only were HAPIC activists asserting their right to hear both sides of an issue, but also their ideas about citizenship were tied to that right. Thus, attending a mock trial or any HAPIC event was a chance for Hyde Park activists to enact their rights as citizens and to participate in American democracy. Insisting on hearing both sides of the issue underscores HAPIC activists' desire to establish inclusive organizing practices that allow all voices to be heard and all those who want to participate to do so. Here again, their historical experiences of exclusion underlie activists' present-day priorities. Moreover, by arguing for H.'s right to be heard, HAPIC activists asserted their own rights to evaluate all aspects of an issue and to make their own decisions on how to act. Thus, they resisted professionals' attempts to dictate how they should react to MOX.

In his closing remarks, Utley delivered an expansive message meant to reunite the trial's audience and participants:

> I don't care if you're white, blue, pink, green, I could care less. Only thing I care about is those who fear God. If you don't fear God, you'll do anything. Whoever your God may be. I'm not going to stand here tonight—I'm not going to preach to y'all. [Audience laughs.] I want

> you to understand that it is more than just the fuel mixing. We're talking about people. We're talking about a lot of people. I want to make sure that this court leaves here this afternoon with an understanding that you have some new information. And I want you to go back and think about the information. Pro or con, but think about it. But make your right decision on your thinking. And I'm going to tell you always to refer to someone that's greater than you to help lead you. And I'm going to close with that by saying that we can make a difference not only in this play court, but in real life, and that is up to each of you. And bless you as you go on your way, and I pray that God will give you the energy to stand for what is right. Amen.

Utley's statement that fuel mixing was about "people" again brings the issue out of the realm of obtuse technology and back to the immediate concerns of Hyde Park residents.[42] By ambiguously describing those people as "white, blue, pink, [or] green" but united by a belief in a higher spiritual power, he also sends a multiracial message and appeals to all audience members. Moreover, deferring to this ambiguous higher power directed activists toward "right" actions and nourished their activism. Thus, for Utley and other HAPIC members, their power as activists is derived from a higher spirituality that gives them "the energy to stand for what is right."

Importantly, Utley leaves the definition of "what is right" up to each individual. As noted earlier, even during its quiescent periods, HAPIC allowed for discourse among members, thus solidifying its bonds as an organization and maintaining its democratic ideals. Here, in a less quiescent period, that emphasis on equal participation extended as far as HAPIC's potential foes. In other words, if each person were going to be able to determine what was right for him- or herself, it was essential to hear *all* sides of an issue. For, as HAPIC activists recognized (and asserted), true access to information, and to democracy, travels along multiple routes.

Conclusion

Each time a group of mainstream environmentalists visited Hyde Park, activists drove them around the neighborhood on a "toxic tour." These environmentalists got a good look not only at each polluter but also at

Hyde Park's drug dealers and the deteriorating houses whose residents could not afford repairs. Their alliances with mainstream professional activists presented HAPIC activists with opportunities to gain more support for their neighborhood's social and ecological problems.

To smooth over tensions that might arise from race and class differences, HAPIC activists used strategic, multiracial language that deemphasized the African American aspects of their situation. However, for them, multiracialism was not merely a way to gloss over potentially sticky issues, and diversity organizing meant more than simply acknowledging their cultural differences with mainstream activists. Rather, it also presented an opportunity to *redress* historical exclusions from white middle-class society and to insert activists' own ideas about democratic participation into mainstream discourses. Moreover, in resisting the notion that the path to empowerment is paved with computer technology, HAPIC activists questioned certain contemporary liberal assumptions about the meaning of public participation and how best to achieve it in the twenty-first century. In other words, they recognized that computer technology, particularly the Internet, does not necessarily offer a universal, magical empowerment potion. As anthropologist Gustavo Lins Ribeiro aptly states,

> It is true that diffusion of information is positively correlated to democratization of access to power. However, if we take into consideration that books, publications, CDs, and the mass media have destroyed neither profound existing social inequalities, nor the abuse of power, we can predict that computer networks will come no closer to representing a true libertarian panacea.[43]

Although a number of scholars have paid particular attention to the degree to which the environmental justice movement has inserted the needs of minorities into mainstream environmental agendas, we have few accounts of how environmental justice activists might also be altering mainstream discourses and practices. Indeed, a good number of social movement studies that examine cross-race and cross-class coalitions tend to assume that more dominant groups hold sway over less dominant groups.[44] Examining such power dynamics is of course critical, but we must not forget the agency of less dominant groups. Rather than completely altering their organizing practices to accommodate the ideas of professionals, HAPIC activists were able to maintain and assert some of their organizing traditions and frameworks for environmental activism. For them, empowerment entailed not just relief from environmental contamination but also resisting being folded into white middle-class ways of organizing and deciding for themselves how to define "right" actions.

Sociologist Francesca Polletta argues that in participatory democracy, the point of deliberation is not unanimity but *discourse*, or making each person's reasoning understandable.[45] By embracing and fostering dissension, even from their potential foes, HAPIC activists ensured that their movement truly lived up to its democratic ideals. In other words, for these African Americans, who have historically been denied their right to partake in the American democratic system, ensuring that all voices were heard was essential to their vision of environmental justice—especially if it is to be the civil, or the human, rights movement of the new millennium.

Seven

No Progress without Struggle

"End it on a positive note," said Arthur Smith when I called him in despair, after staring at my blank computer screen for what seemed like an entire day. I had sat down that morning to write the conclusions to nearly six years of research and writing, not to mention a struggle that is very much ongoing. Synthesis was proving more difficult than I had imagined. "Right," I murmured, thinking of all the ways that my account might portray Hyde Park's quest for environmental justice as hopeless, foolhardy, or depressing—Goliath besting David time and again.

Still, I could see Smith's point. On the one hand, thanks to the 1999 EPA Brownfields grant, the Goldberg scrap metal yard had been assessed and cleaned. Gone were the towers of tires that inauspiciously marked Hyde Park's entrance. Gone, too, was the creaky old shed turned office where contractors found buckets, pails, and jars spilling over with elemental mercury (collected by one of the scrap yard's owners). In fact, once they realized the extent of cleanup needed, the EPA asked the Georgia EPD for assistance. In an almost unprecedented cooperative effort, the two agencies removed 20,000 tons of surface waste, including 6,770 tons of miscellaneous debris, 12,000 tons of hazardous lead-contaminated soil, and 181 tons of mercury-contaminated debris.[1]

The EPD's contractor also hired and trained three Hyde Park residents to help with the cleanup operations; later, he hired one of them to work on another cleanup nearby. In February 2003, HAPIC won a second Brownfield grant for a site across the road. Even more exciting, this grant would test properties on the sites' perimeters and then install monitoring wells in certain people's yards. The EPA has pronounced the SWP site, as well as the lead-contaminated ditch across from Jenkins Elementary School, "clean." And in the summer of 2004, Hyde Park residents convinced city council members to revoke the permit for the new recycling facility that had set up shop across from the former scrap

Drums containing mercury-contaminated debris removed from Goldberg Brothers scrap metal yard, 2000. Photo courtesy of American Environmental and Construction Services, Inc.

Scrap yard after cleanup, 2003. Photo by author.

yard. Mayor Young promised that no more unwanted land uses would be slated for Hyde Park.

On the other hand, it was hard not to think about how the Brownfield project's preliminary investigations had revealed significant levels of arsenic, cadmium, lead, chromium, mercury, PCBs, and petroleum hydrocarbons on the edges of the Goldberg site—a stone's throw from the yards of David Jackson, Hattie Elam, and other residents.[2] I also thought of the metal signs planted throughout the neighborhood warning children not to play in the ditches, some of which are now so faded as to be barely readable. In 1998, the ATSDR reiterated its findings that Hyde Park residents faced no current health dangers, but every time I visit the neighborhood, at least one or two people are being "funeralized." Like many people in Hyde Park, I wonder whether these deaths are brought on by decades of exposure to contamination. I have also considered how in 2000 a federal judge ruled that 158 out of 197 plaintiffs in the case against SWP were ineligible for the class action suit due to "insufficient evidence." Shortly thereafter, the case was "statistically" closed (meaning that it is no longer officially an active case, although either party can petition to reopen it). Finally, although HAPIC activists' recent victory against the new recycling center was certainly significant, promises to remove the landfill at the end of Carolina Road (which residents complained smelled of gasoline) had gone unfulfilled. It was difficult not to question how these facilities had gotten through the permitting process in the first place. How many fires would Hyde Park residents need to extinguish as, every few years, local officials conveniently forgot about their situation?

"Well, how do you mean?" I questioned Smith, always encouraged—and intrigued—by his unrelenting optimism. He replied,

> Too often in the struggle, there's no victories. To get people attached to the movement, you've got to celebrate the victories. The Brownfields grant is to me the end of the rainbow. For a low-income community like ours to know that we are part of the federal government. . . . We've been blessed as a community that so many people—Augusta State University, Paine College, and Clark Atlanta University—understand our pain and our suffering and also want to see relief [for us]. At first, people said that we were on the wrong side of the tracks, that we were gangsters. Who would have thought the federal government, the EPA, and the EPD would be sitting at the table working on a place like Hyde

> Park? So much pain, so much abuse of a community. But with the impact of all of this [contamination], we still carry on a somewhat normal and decent life.

As usual, Smith's speech hit on many of the subjects central to this study. HAPIC's successes and disappointments alike demonstrate some of the ways that African Americans adapt past legacies to contemporary challenges as they struggle to make democratic ideals match their lived experiences.

Indeed, as Smith says, the issue of connecting to governmental organizations, activists, and people in general was a recurring theme in HAPIC's goals and strategies. For example, an HAPIC flyer from the early 1990s reads,

> Even though we have come a long way, we are sometimes made to feel as a "Forgotten Neighborhood." . . . As tax payers and voting citizens of the United States, we should not be forgotten by anyone.

As Terence Dicks recently said, "We're on the map now—they know we're out there." Yet, recent near misses with the permit department attest to the fact that staying on the map requires great effort and persistence.[3] For Hyde Park residents, the pursuit of environmental justice entails putting democratic ideals into practice and integrating neighborhood concerns into governmental agendas. In other words, pursuing environmental justice presents a new opportunity to pursue more traditional goals of democracy and inclusion, or to simply become "part of the government."

Martin Luther King Jr. wrote in a Birmingham jail, "The passage of time alone does not bring change." For change to occur in even incremental ways, people must strive every day to push the barriers that confine their opportunities and to alter the status quo. Over a thirty-year period, this has been the goal of one group of neighbors who have worked steadily to improve the conditions in which they live. Developing a greater understanding of those three decades of struggle—the obstacles residents have faced, the strategies they have used to overcome those obstacles, how they conceptualize and articulate the things that have happened to them, and the things that they wish to make happen —tells us how discriminated-against people can work together successfully to make a difference in their lives and the lives of those around them. For, as Arthur Smith states, Hyde Park's victories lie in many places.

"A Tale Twice Told":[4] Black Activism in the Post–Civil Rights Era

The past has left a twofold legacy for the residents of Hyde Park. First, they have inherited a legacy of slavery, sharecropping, Jim Crow, and institutional barriers to economic opportunity and physical health. In many ways, the story of Hyde Park is the story of a neighborhood in dire straits: every day, residents face poverty, unemployment, drug selling, and pollution. But, second, these residents have inherited strong traditions of successful struggle, forbearance, and community solidarity. Thus, people living in Hyde Park are armed with the faith they find in religion, the memories of past activist victories, and the foundation of a close-knit community that has pulled together many times in the past.

Although many of the hurdles Hyde Park residents face have a clear-cut relation to their being African Americans, in the post–civil rights United States, "race" constitutes an especially elusive social category. The disavowal of a biological basis for race and the outlawing of racial discrimination have led us to celebrate multiculturalism, but they have also made it more difficult to discuss both everyday and structural instances of racism. Nonetheless, in 2004 we still find that fewer than 50 percent of black families own their own homes, compared to more than 70 percent of whites; a black person's average jail sentence is six months longer than a white's for the same crime; and teachers with less then three years experience teach in minority schools at twice the rate that they teach in white schools.[5] Such statistics indicate that racial discrimination in the twenty-first century is an ongoing yet increasingly complicated tale.

Attempts to make sense of these conditions have led to blame-the-victim stereotypes, which posit that if separate races do not actually exist, then the reasons for social inequalities lie in people's behaviors, attitudes, and culture. In this view, equal opportunity must be there for the taking, and if people do not realize it, it is their own fault. Often these ideas are mapped onto neighborhoods themselves. Places like Hyde Park are seen not only to breed maladaptive behavior but also to lack social capital, or the social networks that lead to local power. But Hyde Park shows us that many such neighborhoods are actually close-knit, highly organized, heterogeneous places, with extensive social networks. We also know from this case that many residents who choose to move out of these neighborhoods return often, either to socialize or to work on neighborhood problems. What keeps Hyde Park residents from accessing the resources they need are persistent structural barriers to racial equality. Indeed, Hyde Park illustrates that our ideas and discourses about concepts such as race have concrete consequences for people's everyday lives. It is no wonder, then, that the concept of race continues to have great significance for, if not to be an organizing principle of, social life.

An Unfair Share: The Environment, Race, and Justice

Whatever the environment means to different people in different cultures, environmental burdens are unevenly distributed around the globe, and people of color and poor people are on the front lines of poisonous exposures. However, because the environment emerged in public discourse as a white, middle-class concern (both in the United States and abroad), many people of color did not initially name their environmental problems as such, despite the fact that these people have *always* cared deeply about natural resources.

Grassroots environmental justice activists in cities, villages, and rural settlements throughout the world are struggling to address the roots of environmental racism. Some of these efforts mirror HAPIC strategies, and many of them tackle the topics touched on in this book. For instance, activists in the United States are working to establish holistic profiles of contaminated communities that account for the many ways in which people are exposed to toxic chemicals throughout their lifetimes. They are also building participatory and community-driven research models. More abstractly, these efforts show how grassroots activists are challenging the notion that only empirical science is accurate or reliable, and asserting the value of experiential data. Like the people in this book, they recognize that science is fallible and that facts can always be disputed with more facts. Thus, they question whose representation of the "truth" is privileged and whose is silenced. Finally, environmental justice activists point out that having evenly distributed environmental burdens does not equal environmental justice—it only equals more (if different) sick people. Thus, they are working to reduce the production of waste in the first place.

Grassroots environmental justice activists worldwide have also formed important partnerships with mainstream environmental groups. This book shows how such partnerships are facilitated through the flexibility of the "environment" as an organizing narrative. In other words, Hyde Park activists came to see the environment as encompassing clean air, water, and soil, as well as crime, violence, poor educations, and unemployment. Thus, they made it a locally salient issue. But they also expanded their environmental definitions in a second way—as affecting all races, colors, and creeds alike. This second expansion then enabled them to tap into the resources of the more economically and politically powerful mainstream environmental movement. As I write this in the late summer of 2004, for example, HAPIC is hosting a group of Russian antinuclear activists. No doubt they will escort the Russians on a "toxic tour," showing them the former scrap metal yard and wood-processing plant, as well as the dilapidated houses, the elderly folks who sit on their porches to escape the sweltering heat of their non-air-conditioned homes, and the Utley Center, now badly in need of repair. In this way, HAPIC activists will not only contribute to international struggles for environmental justice but also encourage their new colleagues to conceive of the environment as both a social and an ecological issue.

No Progress without Struggle: Social Movement Organizing and "The Long Haul"

Environmental justice teaches us that influences between social movements do not always move in a top-down direction, with a more powerful group overshadowing its less powerful counterparts. For instance, by integrating religion into their environmental activism, HAPIC members upheld black organizing traditions and inserted their own values into the secular practices of mainstream environmentalists. Moreover, in questioning mainstream assumptions about the role of technology in social movement organizing, as well as the meaning of participation, HAPIC members asserted their own ideas about how to achieve power and participatory democracy. HAPIC members thus not only extended their networks to include more influential groups but also taught the groups with which they networked alternative ap-proaches to environmental organizing.

Social movements are dynamic on both large and small scales. Local participation in HAPIC varied widely over the years, as did the opinions of the organization's members. Some residents believed that relocation was the only solution to their problems and continued to devote ample energies to fighting for it. Others did not want to move and preferred to search for ways to clean the neighborhood, no matter how long it took. Still others had grown tired of fighting what seemed like a losing battle and no longer believed that contamination constituted a great threat to their health. In the late 1990s, as their lawsuit dragged on and little help came from governmental entities, HAPIC's active membership dwindled. Since the Brownfields grant took effect in 2000, some residents have warily returned to the organization and continue gradually to step up their involvement. In 2003, for example, a busload of HAPIC members attended a national Brownfields conference in Charlotte, North Carolina.

Throughout the years, regardless of fluctuating attendance or sporadic meeting schedules, Hyde Park residents have engaged in "quiescent politics," sustaining a neighborhood solidarity, organizing infrastructure, and maintaining an infectious activist consciousness. Coming together meant that HAPIC members recalled past successes and hardships. They also commiserated over the difficulties of contemporary life in Hyde Park, discussed ways to contest their situation, and sometimes complained about their leaders. By sharing these memories and complaints, even if they did not lead to concerted overt actions, HAPIC meetings served an important purpose: through them, members enacted the democratic ideals of equal participation and free expression upon which they founded their organization. Such enactments not only bolstered an organizational identity but also had concrete outcomes. For instance, in the summer of 2002, as trucks began dumping soil on the edge of Carolina Road, and then again in 2004, as residents watched the recycling center take its place across from the recently cleared scrap yard, people immediately began holding meetings, assigning tasks, lobbying the city council, and contacting the local press. In short, over three-plus decades, HAPIC activists have never let go of their goals to become "part of the government" and to make democratic ideals match their lived experiences.

HAPIC shows us, then, that progress toward social change might be halting or slow, or sometimes might even take a few steps backward, but there *is* progress if you look for it. Sometimes it is as obvious as winning a Brownfields grant. Other times, you have to dig a little deeper. If you do, you can find victory in the simple sharing of memories of past victories. Or you can find progress in the very fact that a small neighborhood like Hyde Park has formed strong, lasting partnerships with more powerful organizations and, through those partnerships, has influenced the practices of those more powerful organizations. Or, that even though they need to restage the same battle every few years, Hyde Park residents *do* periodically show the rest of Augusta that their toxic waste and unwanted land uses go somewhere, and often they affect real people, who are struggling to live out very real hopes and dreams.

Just over the Hill

Although this book has focused on the meaning of the environment, race, and a civil rights–era legacy for African Americans, by extension, it also has much to tell us about the meaning of civil rights and the environment for *all* Americans.[6] It has been the privilege of white people in the United States to not have to think about the daily indignities of being charged too much for a car loan, or being steered toward one neighborhood over another simply because of skin color. It has also been middle- and upper-income whites' privilege to assume that a "free" market will distribute environmental burdens equally among all social groups, or that waste somehow takes care of itself—as long as it is unseen, it remains unproblematic. However, even if it is poor, and especially minority, groups who are in the direct line of fire when it comes to environmental contamination, the poisoning of the environment—like racism—endangers all of us. If left unattended, the effects of the arsenic, cadmium, chromium, lead, and mercury that have been allowed to contaminate Hyde Park's soil, water, and air will eventually radiate throughout Augusta and, perhaps, beyond. In short, just as racism diminishes our nation's economic, political, and social potential, pollution is "democratic";[7] environmental injustice anywhere is truly a threat to health and well-being everywhere. Or, as Arthur Smith put it, "[Hyde Park's] train tracks aren't going to stop toxic chemicals. They're going to sneak up to bite you, too."

Appendix A: Methods

October 7, 1997
VIA FACSIMILE
Dear Mr. Utley,
Thank you so much for taking the time to speak with me on the telephone yesterday. I am sorry for your troubles in Augusta, and I hope you and your neighbors will get some kind of relief in the near future. I also hope that I can be of some help to you. I became interested in environmental justice as an activist, and I now want to combine that activism with my academic work. My research will focus on one grassroots environmental justice group whose ranks I hope to join. From what you explained to me yesterday, I think that your group would be an excellent place for me to learn about environmental justice organizing. I intend to participate in your group's activities as a volunteer, lending whatever assistance I can. . . . I am committed to the cause of environmental justice and want to emphasize my intention to participate as an activist as well as to conduct research as an anthropologist. . . .

In the fall of 1997, I sent Charles Utley the preceding letter to follow up a telephone conversation. Although our call was rushed because Utley had to run off to coach his school's football team, that talk marked the last in a series of phone calls I had made that autumn in my search for a field site.

In 1994, I had conducted research with a multiethnic group of minority activists in Brooklyn, New York, who were opposing the installation of an incinerator in their already polluted neighborhood. This time, I decided to move southward and explore environmental justice activism in the region where it began in the late 1980s. However, I had no contacts with southern environmental justice activists. I received the names of potential groups with which I could work through various sources, but unsurprisingly, until the day I phoned Utley, no one had even let me present my spiel.

With characteristic patience, Utley listened long enough to hear and digest my offer: I would come to and live in Augusta for a year and act as a nonpaid staff member of his organization. My job description would be created by HAPIC's activists, and I would agree to perform whatever duties they required. In exchange, activists would allow me to sit in on meetings and events and have access to their files. HAPIC had no staff, relying entirely on the volunteered time of its membership. Even so, HAPIC had received several operating grants and become well known among environmental justice activists across the nation, even traveling to Washington, D.C., to testify before members of Congress and other environmentally related subcommittees. Wishing to continue and enhance his organizations' successes, Utley recognized in my offer a potentially beneficial situation. After consulting with HAPIC's board members, he agreed to have me come to Hyde Park.

When I arrived in the fall of 1998, I had lunch at a popular southern-style cafeteria with Utley and Melvin Stewart, HAPIC's treasurer. The two men—longtime friends and colleagues—suggested that I set up base at the Mary Utley Community Center, a neighborhood center from which HAPIC operated. The Utley Center also housed offices of the Economic Opportunity Authority (EOA), a local nonprofit that assisted low-income Augustans with electric, gas, and rent bills. The combination of the EOA and HAPIC drew many residents to the Utley Center on a daily basis, making it a superb place from which to observe and participate in neighborhood goings-on.

Utley and Stewart also requested that I start an after-school tutoring program for Hyde Park children in grades K–12 that would have some kind of environmental education component, and that I help HAPIC get 501(c)3 status, making it an "official" nonprofit. I hesitantly suggested that I might also build it a Web site. "A Web site!" Utley exclaimed and looked to Stewart.[1] Both grinned with the wide, easy smiles I would soon come to know well. Thus began my fourteen months of participant observation with the Hyde and Aragon Park Improvement Committee—with an extra emphasis on "participation."

Participant Observation and Reciprocity

While reciprocity (as well as activism) has always been a tradition in anthropological fieldwork, it has remained a largely undiscussed topic for fear that the subjectivity it implies will compromise anthropology's legitimacy as a serious social science.[2] In the past twenty or so years, many anthropologists have recognized the limitations of trying to achieve "objectivity," acknowledging that subjectivity enters into and complicates *any* research project. Anthropologist Renato Rosaldo argues that rather than seeing these subjective positions as compromising, we might recognize that objective stances themselves are "neither more nor less valid than those of more engaged, yet equally perceptive, knowledgeable social actors."[3] Taking Rosaldo's argument one step further, I suggest that we view the ethnographer's project on a continuum of engagement. The ethnographer then situates herself on that continuum and analyzes her research from that place. For example, my commitments to environmental justice activism position me on a somewhat far end of the engaged spectrum.

But that position has only partly to do with my own personal commitments to environmental justice activism. To a large degree, it also stems from strong feelings about finding a clear-cut way to "repay" the residents of Hyde Park and the activists of HAPIC for allowing me to do research with them. Volunteering full-time for HAPIC not only enabled this "repayment" but also benefited my fieldwork method in myriad ways. First, I was an "outsider" on many levels. During my first two weeks at the Utley Center, I was known mainly as "the white lady."[4] While Hyde Park residents had seen other white activists come into their neighborhood, the daily presence of a white person, and a northeasterner, definitely sparked questions, curiosity, and skepticism.[5] Despite my repeated explanations, no one was quite sure why I was there, what I planned to do, or how long I planned to stay. Yet, while they did not necessarily believe that their neighborhood needed an anthropologist, nearly all Hyde Park residents agreed that it needed a tutoring program.

As soon as I began the after-school tutoring program, parents would stop in at the Utley Center when they picked up their children to meet and greet me. I held several meetings with parents to discuss their children's progress, and in those meetings I had the chance to ask neighborhood adults about local contamination and other conditions. I also observed parent-child and neighbor-child interactions, giving me insights into the close-knit flavor of Hyde Park and the degree to which "mainstream" ideas about child rearing and education are very much a part of their lives.

As the months of fieldwork wore on, I became more engrossed in the daily business of HAPIC, and activists began to trust me with increased responsibilities and access to the organization's files and financial records. About three-fourths of the way into the fieldwork, my typical day at the Utley Center might include working on a grant proposal for the organization, working on a flyer for an upcoming community meeting, driving someone to do an errand,[6] answering requests from funders for reports and other information, sorting through files, conducting an interview with an HAPIC member, chatting with residents as they dropped into the center, and preparing for evening programs. By the end of my fieldwork, I had written four grant proposals for HAPIC (three were successful), created a Web site, organized after-school and summer programs for youth (including environmental education, tutoring, field trips, and outdoor sports), helped plan to plan numerous community meetings and two community cleanup days, and organized adult computer training courses.

I eventually found that participating in HAPIC aided my research, and sometimes my research aided my participation in HAPIC. In organizing a community cleanup or meeting, for example, I called upon certain HAPIC members to help me pass out flyers. In some cases, contacting people for this purpose led to an interview. In other cases, this process worked in reverse—I met people at community meetings, asked them for an interview, and during it managed to corral them into helping pass out flyers in the future. Writing grants necessitated developing a written history of HAPIC, which I later used in chapters 4 and 5 of this book. Perhaps most important, my responsibilities to HAPIC kept me at the Utley Center on a daily basis. I thus got to know residents who came in and out of the center and participated in countless informal conversations and discussions about their lives, the lives of their neighbors, and local and national politics.[7] Some of these encounters led to invitations into people's homes for cookouts and birthday parties, or to a Sunday service at their church, all of which enriched my knowledge of social life in Hyde Park.

Another example of symbiosis between my duties as an HAPIC activist and as a fieldworker was my regular presence at a number of local meetings, including Augusta Clean and Beautiful (a group of neighborhood activist organizations by the Augusta–Richmond County government to work on cleaning up Augusta's low-income neighborhoods); the Transportation Leadership Development Initiative (a group organizing to improve public transportation in the county); the Augusta Neighborhood Associations Alliance (an alliance of neighborhood association presidents); and the Jenkins Elementary School PTA. My participation at meetings helped out when HAPIC activists had prior engagements or work commitments. In addition, I gained valuable insights into how local politics worked in Augusta, and the ways in which people understood, discussed, and tried to change it. In the case of the Jenkins PTA meetings, I developed a clear idea of what issues concerned Hyde Park parents and what inspired them to come out and attend community meetings.

Defining my commitments to reciprocity was not always a clear-cut process. On a few occasions, for example, Charles Utley asked me to go in his stead to fairly critical meetings such as the Committee for African American Environmental Justice's Leadership Training Meeting in Washington, D.C., or a meeting with Mayor Young to discuss HAPIC's writing a federal grant proposal on behalf of Augusta–Richmond County. One day, Utley also asked me to compose a mission statement for HAPIC. These kinds of requests caused me some consternation. On the one hand, I had promised to do whatever was asked of me. On the other, involving myself in HAPIC's affairs to the extent that I represented it to the mayor or wrote its mission statement, seemed to overstep even my heavily activist-oriented bounds as a researcher. In the end, I negotiated such situations by (as much as possible) trying to recapture activists' own words and deferring to longer-term members of the group. Finally, my extensive involvement in HAPIC led to some mixed personal emotions. By becoming an integral part of the organization, I went through many of the same ups and downs as its members, as together we encountered both the successes and the failures that come with the daily practice of community organizing.

Interviews

Whether passing out flyers, distributing bags of food at a community cleanup, delivering kids to parents after a program, or hanging out at the local bar, I met at least three-fourths of the 250 adults living in Hyde Park. By attending numerous meetings and events, I also met most of the people who regularly worked with HAPIC, even if they did not live in Hyde Park. Eventually, I established a group of approximately twenty-five people who were key providers of information and with whom I conducted formal interviews. Of those, I got to know about seventeen or eighteen fairly well, and I have interviewed most of them two or three times over the past five years. Some of their stories make up my chapter prologues.

I chose interviewees based on how long they had lived in Hyde Park and/or been part of HAPIC and the degree to which they represented some aspect of organizing that I wanted to understand. Approximately half of the people I interviewed were currently active participants or leaders in HAPIC. (I define "active" participants as those who attended almost every meeting and who helped plan certain events and/or pass out flyers.) Another thirteen had been active in HAPIC in the past but no longer attended meetings on a regular basis or assisted in planning or executing organizational activities. These people also included professionals who had volunteered to help HAPIC in some way—for example, its lawyers, a local professor, and several other activists who were allied with outside organizations. Almost all those I interviewed were African American and had spent most of their lives in Hyde Park. Thanks to my research assistant, Michelle (who grew up in Hyde Park), I was able to meet and interview a number of "old-timers," people in their late sixties, seventies, or eighties who provided detailed oral accounts of neighborhood history.

My interview style was "semidirected." That is, I asked specific questions on such topics as changes in Hyde Park and HAPIC over the past thirty years; the meaning of the environment; attitudes toward neighborhood activism and activists; and the meaning of racism. These queries included "small" questions (e.g., "In what specific organizational activities have you engaged?") as well as "big" questions (e.g., "What does environmental justice mean to you?"). In addition, I took cues from my interviewees, following up on their comments with questions inspired by the course of conversation. Interviewees chose the locations for their interviews, with most preferring to have them in their homes but some at the Utley Center. With activists who did not live in Augusta, I generally conducted interviews over the phone.

With permission, I tape-recorded all in-person interviews and took notes on them. To analyze these interviews, I first transcribed and then coded and examined them for recurring themes. For instance, I looked at the words people used to describe the environment, how they explained the reasons for the contamination of their neighborhood, what memories were most prominent when they discussed the "old days," what they thought were HAPIC's most and least successful events, and other narrative patterns that deepened my insights into how people formulate understandings of political action, racial identity, and the environment.

Names

Most of the people I interviewed are activists and have been for much of their adult lives. They are proud of their accomplishments. When I asked whether they wanted anonymity, most *requested* that I use their real names. As Arthur Smith said, "If you can't stand behind what you've said, you lose credibility as an activist." Consequently, as much as possible, I have used people's real names when they have requested it. For those who did not want to be identified, I have used pseudonyms and changed some of their biographical data. Nearly everyone interviewed has reviewed the material in which they are mentioned and given me final, written permission to publish it.[8]

Survey

Upon arriving in Augusta, I was fortunate to become acquainted with faculty members of the Augusta State University (ASU) Department of Sociology. In conjunction with ASU's student sociology club and a class in survey methods, the sociology department had decided to conduct a Hyde Park neighborhood survey in the fall of 1998. I, too, had planned a survey, and to spare Hyde Park residents the annoyance of answering two surveys within several months, I worked cooperatively with the ASU researchers on theirs. The survey consisted of forty-one questions designed to measure residents' attitudes and opinions about neighborhood concerns and behaviors, as well as environmental issues and involvement in HAPIC. In addition, it tested theoretical models for explaining participation in grassroots community organizations.[9] ASU faculty members allowed me to preview and have some input into survey questions, a good number of which matched my own research questions. They then organized groups of students to go door-to-door in the neighborhood to distribute initial questionnaires and obtain results in person. Later on, students completed follow-up interviews as needed. The survey (conducted over four consecutive weekends) resulted in completed questionnaires from 176 individuals, or approximately two-thirds of Hyde Park's adult population. In its testing of sociological models, the survey diverged from my more ethnographic approach to social movement study. I have relied on it primarily to supply demographic information on Hyde Park (since the area does not constitute its own census tract, reliable demographics are otherwise hard to come by). In a few cases, I have also used it to record residents' attitudes toward HAPIC and neighborhood problems.

Archival Data

Finally, I supplemented my work with individuals and groups by paying close attention to representations of relevant issues in the local media. For example, I clipped newspaper articles from Augusta's four newspapers—the *Augusta Chronicle,* the *Metropolitan Spirit,* the *Metro Courier,* and the *Augusta Focus* (the latter two are owned and operated by African Americans)—that addressed issues pertaining to race and environment. To gather historical material and statistical data, I used materials on file at the ASU library and census data. I also spent one long, fruitful afternoon in the archives of the Richmond County Historical Museum and several slightly more frustrating afternoons at Augusta–Richmond County's records department.

Conclusion

Conducting fieldwork in your own country has both advantages and disadvantages. One of the major advantages is that the people with whom you develop the close ties that fieldwork fosters are never far away. Since leaving Augusta in October 1999, I have returned to Hyde Park on eight separate occasions. During these visits, I divide my time between follow-up interviews, attending any HAPIC events that happen while I am there (this frequently and deliberately occurs), and visiting people just to catch up. Twice, I also conducted community seminars as part of the Brownfield grant program. In the months between visits, I speak regularly on the telephone with several people. That allows me to stay in touch with individuals and neighborhood goings-on, and continue some of my participation in HAPIC business, especially grant writing and maintaining the Web site. Continuing my relationships with the people of Hyde Park has enabled me to resolve research questions as they arise, to fill in gaps as they appear, to stay somewhat active in HAPIC, and, most important, to maintain valuable friendships.

Appendix B: Getting Involved

The following is a selected list of sources for environmental justice information and organizing activities:

Action for Community and Ecology in the Regions of Central America (ACERCA)
A project of Action for Social and Ecological Justice (ASEJ), ACERCA supports indigenous peoples and environments in southern Mexico and Central America, using those regions as a lens on the global economy.
PO Box 57, Burlington, VT 05402
802-863-0571
www.asej.org

African American Environmental Justice Action Network (AAEJAN)
The mission of the AAEJAN is to address the disproportionate impact that pollution and toxins have on African American communities in the South and to work with others to achieve environmental justice and a better quality of life for all.
PO Box 150196, Atlanta, GA 30315
404-627-7790
aaejan1@juno.com

Asian Pacific Environmental Network (APEN)
The mission of APEN is to unify and empower the Asian Pacific Islander community to achieve multicultural environmental justice.
310 Eighth St., Suite 309, Oakland, CA 94607
510-834-8920
510-834-8926 fax
apen@apen4ej.org
www.apen4ej.org

Association of Community Organizations for Reform Now (ACORN)
ACORN is the nation's largest community organization of low- and moderate-income families, with more than 175,000 member families organized into 850 neighborhood chapters in more than seventy cities across the country.
88 Third Avenue, Brooklyn, NY 11217
718-246-7900
skest@acorn.org
www.acorn.org

Center for Health, Environment and Justice (CHEJ)
CHEJ was founded in 1981 by Lois Gibbs, a leader of the campaign at Love Canal, and was formerly called the Citizens Clearinghouse for Hazardous Waste. Since then, CHEJ has expanded its programs to match the expanding concerns of grassroots environmental groups.
PO Box 6806, Falls Church, VA 22040
703-237-2249
chej@chej.org
www.chej.org

Center for Policy Alternatives (CPA)
The CPA is the nation's leading nonpartisan progressive public policy organization serving state legislators. CPA strengthens the capacity of state legislators to lead and achieve progressive change. CPA staff connect legislators with advocates and experts to develop public policy solutions, share information and program models, and enhance policy advocacy skills.
1875 Connecticut Ave. NW, Suite 710, Washington, DC 20009
202-387-6030
info@cfpa.org
www.stateaction.org

Center for Third World Organizing (CTWO)
CTWO, pronounced "C-2," is a racial justice organization dedicated to building a social justice movement led by people of color. CTWO is a twenty-four-year-old training and resource center that promotes and sustains direct action organizing in communities of color in the United States. CTWO's programs include training of new and experienced organizers, including the well-known Movement Activist Apprenticeship

Program (MAAP); establishing model multiracial community organizations; and building an active network of organizations and activists of color to achieve racial justice in its fullest dimensions.
1218 E. Twenty-first St., Oakland, CA 94606
510-533-7583
ctwo@ctwo.org
www.ctwo.org

Children's Health Environment Coalition (CHEC)
CHEC is a national nonprofit organization dedicated to educating the public, specifically parents and caregivers, about environmental toxins that affect children's health.
PO Box 1540, Princeton, NJ 08542
609-252-1915
www.checnet.org

Concerned Citizens of Rutherford County, NC (CCRC)
CCRC is a grassroots community-based group located in Rutherford County, North Carolina. Since its inception in June 1995, CCRC has grown from a locally based organization to serving as a voice for communities throughout the Southeast and Appalachian regions dealing with unsustainable forestry practices. CCRC mentors community groups in Alabama, Kentucky, Missouri, North Carolina, Pennsylvania, South Carolina, Tennessee, Virginia, and West Virginia.
PO Box 623, Rutherfordton, NC 28139-0623
828-287-4429
ccrc@rfci.net
www.ccrcnc.org

Defense Depot of Memphis Tennessee–Concerned Citizens Committee (DDMT-CCC)
DDMT-CCC provides communities in need with technical and organizing assistance so that they can defend their families from toxic pollution, and fight back if they discover they have been poisoned or violated in other ways. Efforts include work around reproductive rights, police brutality, youth concerns, environmental justice, and other social justice issues.
1000 S. Cooper, Memphis, TN 38104
901-726-0008
ddccc@bellsouth.net

Environmental Health Coalition (EHC)
EHC organizes and advocates to protect public health and the environment threatened by toxic pollution. It supports broad efforts that create a just society that fosters a healthy and sustainable quality of life.
401 Mile of Cars Way, Suite 310, National City, CA 91950
619-474-0220
ehc@environmentalhealth.org
www.environmentalhealth.org

Environmental Working Group (EWG)
EWG's team of scientists, engineers, policy experts, lawyers, and computer programmers pores over government data, legal documents, scientific studies, and its own laboratory tests to expose threats to health and the environment and to find solutions.
1436 U St. NW, Suite 100, Washington, DC 20009
202-667-6982
www.ewg.org

Indigenous Environmental Network (IEN)
IEN is an alliance of grassroots indigenous peoples whose mission is to protect the sacredness of Mother Earth from contamination and exploitation by strengthening, maintaining, and respecting traditional teachings and natural laws.
PO Box 485, Bemidji, MN 56619
218-751-4967
ien@igc.org
www.ienearth.org

Military Toxics Project (MTP)
MTP's mission is to unite activists, organizations, and communities in the struggle to clean up military pollution, safeguard the transportation of hazardous materials, and advance the development and implementation of preventative solutions to the toxic and radioactive pollution caused by military activities.
PO Box 558, Lewiston, ME 04243-0558
207-783-5091
mtp@miltoxproj.org
www.miltoxproj.org

National Pesticide Information Center (NPIC)
NPIC is a cooperative effort of Oregon State University and the U.S. Environmental Protection Agency. It offers a toll-free telephone service that provides pesticide information to any caller in the United States, Puerto Rico, or the Virgin Islands. NPIC is staffed by highly qualified and trained pesticide specialists who have the toxicology and environmental chemistry education and training needed to provide knowledgeable answers to pesticide questions. It is a source of factual chemical, health, and environmental information about more than six hundred pesticide active ingredients incorporated into more than fifty thousand different products registered for use in the United States since 1947.
Oregon State University, 333 Weniger, Corvallis, OR 97331-6502
1-800-858-7378
www.npic.orst.edu

Northwest Coalition for Alternatives to Pesticides (NCAP)
NCAP is a five-state, grassroots membership organization that promotes sustainable resource management, prevention of pest problems, use of alternatives to pesticides, and the right to be free from pesticide exposure.
PO Box 1393, Eugene, OR 97440
541-344-5044
info@pesticide.org
www.pesticide.org

Office of Management and Budget (OMB) Watch
A nonprofit research and advocacy organization dedicated to promoting government accountability, citizen participation in public policy decisions, and the use of fiscal and regulatory policy to serve the public interest, OMB Watch primarily tracks the operations of the U.S. Office of Management and Budget and is part of the Right to Know (RTK) Network.
1742 Connecticut Ave. NW, Washington, DC 20009
202-234-8494
www.ombwatch.org

Silicon Valley Toxics Coalition (SVTC)
SVTC is a diverse grassroots coalition that engages in research, advocacy, and organizing around the environmental and human health problems caused by the rapid growth of the high-tech electronics industry.

760 N. First St., San Jose, CA 95112
408-287-6707
www.svtc.org

Southwest Network for Environmental and Economic Justice (SNEEJ)
SNEEJ is made up of grassroots, community-based, native, labor, and student groups across the Southwest, West, and border states of Mexico building a multiracial, multicultural, multigenerational, and international movement proactively working for sustainable communities for environmental, economic, social and racial justice.
PO Box 7399, Albuquerque, NM 87194
505-242-0416
info@sneej.org
www.sneej.org (bilingual site, English and Spanish)

State Environmental Leadership Program (SELP)
SELP is an alliance of more than fifty independent, nonprofit, public interest, multi-issue environmental advocacy organizations that focus on state-level policy. The program's goal is to strengthen state environmental movements by enhancing the organizational capacity and policy work of SELP's member organizations through network collaboration.
612 W. Main, No. 302, Madison, WI 53703
608-268-1440
general@selp.org
www.selp.org

Student Environmental Action Coalition (SEAC)
SEAC, pronounced "seek," is a student- and youth-run national network of progressive organizations and individuals whose aim is to uproot environmental injustices through action and education. It defines the environment as including the physical, economic, political, and cultural conditions in which we live. By challenging the power structure that threatens these environmental conditions, SEAC works to create progressive social change on both local and global levels.
PO Box 31909, Philadelphia, PA 19104
215-222-4711
ncc@seac.org
www.seac.org

Women's Environmental and Development Organization (WEDO)
WEDO is an international advocacy organization that seeks to increase the power of women worldwide as policymakers at all levels in governments, institutions, and forums to achieve economic and social justice, a healthy and peaceful planet, and human rights for all.
355 Lexington Ave., 3rd Floor, New York, NY 10017-6603
212-973-0325
wedo@wedo.org
www.wedo.org

Youth Terminating Pollution (YTP)
YTP was established in 1998 to empower youth in the neighborhoods surrounding the former defense depot of Memphis, Tennessee. YTP works on environmental justice issues on a local, national, and global scale.
1000 S. Cooper, Memphis, TN 38104
901-726-0008

Notes

NOTES TO CHAPTER 1

1. Following the general usage of the people in this book, I use the terms "African American" and "black" interchangeably.

2. Children's names have been changed, although those of many adults have not. See appendix A for a detailed explanation.

3. Contamination in this case is loosely defined as the presence of contaminants in air, water, or soil, over the permissible standards set by either federal or state government. In the case of Hyde Park, it is certain that contaminants exist in the neighborhood; I therefore use the term without qualification. However, as I will discuss later in this chapter and many times throughout the book, whether or not contaminants pose a health risk to residents is a highly contested issue.

4. "Superfund" is the popular name for a "fund established by Congress to pay for containment, cleanup, or remediation of abandoned toxic waste sites. The fund is financed by fees paid by toxic waste generators and by cost-recovery from cleanup projects" (highered.mcgraw-hill.com/sites/0070294267/student_viewo/glossary_s-z.html).

5. See Checker 2001, 2002a.

6. Portions of the material contained in this book have appeared in several journal articles and a book chapter that I authored. See Checker 2002a, 2002b, 2004a, 2004b, 2005.

7. HAPIC is so named because the organization includes residents from Hyde Park and Aragon Park, a very small neighborhood across the highway. In the 1970s, when HAPIC was formed, the highway did not yet exist, and the two neighborhoods interacted often; by the late 1990s, Aragon Park residents rarely participated in HAPIC events (this was a controversial topic that is beyond the scope of this chapter). However, Hyde Park activists did not wish to change the name of their organization in the hope that more Aragon Park residents would one day rejoin it.

8. Throughout its sixty-one-year history, Southern Wood employed a variety of wood-processing treatments that contained a range of chemical preservatives,

including creosote, pentachlorophenol (PCP), chromated copper arsenate, and zinc meta-arsenate (see http://www.atsdr.cdc.gov/HAC/PHA/swp/swp_p4b.html).

9. Exactly who made the decision to close the plant is controversial. A report issued by a governor's task force selected to study the health risks of Hyde Park–Virginia Subdivision residents states that the Georgia Environmental Protection Division (EPD) ordered the factory to close and clean up. SWP's attorney, however, has stated that the factory's parent company made the decision to close for a number of reasons, one of which was that soil and groundwater contamination underneath the plant could not be adequately remediated as long as it remained in operation. In either case, at the time of the closing, SWP officials signed a consent order with the EPD, admitting to contamination immediately around the plant.

10. Many Virginia Subdivision residents believed that they had not received enough money to move out of the neighborhood. For that reason, Virginia Subdivision activists joined forces with HAPIC in 1991 and assisted it in finding both legal and extralegal remedies for contamination.

11. In 1920, Colonel Sosthenes Behn founded the International Telephone and Telegraph Corporation (ITT) as a research-driven producer of equipment for telephone and other communication technologies, a role boosted by the company's expansion during World War II. Following the war, ITT continued to expand into a variety of sectors, including automotive, satellite communications, fluid technology, and even Hartford Insurance and Sheraton Hotels (http://www.skillcircle.com/resources/profiles/812.html). In 1968, ITT bought Rayonier, a company specializing in forest products. *Time* magazine, commenting on ITT in 1972, noted that a consumer who was not happy with the company and who wanted to escape its grasp "could not rent an Avis car, buy a Levitt house, sleep in a Sheraton hotel, park in an APCOA garage, use Scott's fertilizer or seed, eat Wonder Bread or Morton's frozen foods. . . . he could not have watched any televised reports of President Nixon's visit to China. . . . he would have had to refuse listing in Who's Who; ITT owns that too" (Geisst 2000). Around that time, ITT fell under national scrutiny when questions arose about its alleged role in plotting to assassinate Chile's new Marxist leader, Salvador Allende (see www.buyandhold.com/bh/en/education/history/2002/itt_pt_2.html).

The corporation bounced back, and at the end of the 1980s, ITT's sales were over $20 billion. In 1993, to streamline its operations, ITT announced it would spin off Rayonier to its shareholders (http://www.endgame.org/dtc/i.html). Today the company specializes in military defense systems and is the world's premier supplier of pumps, systems, and services to move and control water and other fluids. In 2003, sales reached $5.63 billion (http://www.ittind.com/profile/fact.asp).

12. Virginia Subdivision was not an exclusively white neighborhood. In fact, its lawsuit against SWP was spearheaded by two African American residents.

13. ATSDR 1995:6.

14. See ATSDR 1995; Craig 1991a; Governor's Task Force Report 1996.

15. Elevated levels of lead and PCBs, in particular, were mentioned in every report.

16. National Urban League 2004.

17. See Latour 2004 for an excellent synopsis of this conundrum, especially as it relates to environmental issues.

18. For exceptions, see Gregory 1998; Sacks 1994. For historians' accounts of politics and the formation of black identities, see Trotter 1985; Lipsitz 1995 [1988]; Kelley 1990, 1994; Higginbotham 1993.

NOTES TO CHAPTER 2

1. Commission for Racial Justice 1987. A follow-up study found that these numbers had increased by as much as 8 percent by 1993 (Center for Policy Alternatives 1994).

2. Been and Gupta 1997; see also Been 1993.

3. American Lung Association 2004, 2002.

4. Lavelle and Coyle 1992; see also Bullard 2000.

5. Center for Policy Alternatives 1994; see also Institute for Southern Studies 1992.

6. Chavis 1993:3.

7. See Anderton et al. 1994. In Hyde Park's case, both Southern Wood and Babcock and Wilcox (now Thermal Ceramics) were in place well before residents moved in.

8. Massey and Denton 1993:96; see also Cole and Foster 2001:69.

9. Rabin 1990. See also di Leonardo 1998; Quint 1993; Hansell 1993.

10. See, for example, Kirschenman and Neckerman 1991; Newman 1999; Sullivan 1989; Wilson 1996.

11. Williams 1980:71, emphasis mine.

12. Dryzek 1999:278. See also Milton 1995:167.

13. As quoted in Shanklin 1998:675.

14. Adamson, Evans, and Stein 2002:4.

15. For information on these and other U.S. environmental justice movements, see Adamson, Evans, and Stein 2002; Alley, Faupel, and Baily 1995; Bullard 1993, 2000; Cole and Foster 2001; Faber 1998; Moberg 2001; Pellow and Park 2003; Pulido 1996; Roberts and Toffolon-Weiss 2001; Townsend 2000; Williams 2001. For examples of global environmental justice movements, particularly from anthropologists, see Agarwal 1992; Brosius 2001; Bryner 2001; Crumley 2001; Johnston 1994, 1997, 1998; Westra and Wenz 1995. For recent reviews of social science research on environmentalism in general, see Downs 2000; Goldman and Schurman 2000; Little 1999; Schell and Denham 2003; Scoones 1999.

16. Williams 2001:410. See also Crumley 2001; Johnston 1994, 1997, 1998; Westra and Wenz 1995. For full-length environmental ethnographies, see, for example, Alley 2002; Griffith and Valdez-Pizzini 2002; Satterfield 2002

17. Gottlieb 1993:240.

18. Gottlieb 1993:7.

19. Darnovsky 1992:24.

20. Taylor 1992:53.

21. D. Taylor 1989; see also Bullard 2000; Darnovsky 1992.

22. Gottlieb 1993:77; see also Scheffer 1991.

23. Harvey 1996:395; Gottlieb 1993:246–249.

24. The Love Canal residential area was adjacent to a chemical and plastics plant. After two decades of health complaints from residents, the EPA found that leaking gases containing pesticides, chlorobenzenes, and dioxins were causing alarming rates of cancer.

25. Bosso 1991; Brosius 1999; Bryner 2001; Szasz 1994.

26. See Bryant 1995a; Bullard 1993, 2000; Harvey 1996; Novotny 1995, 1998.

27. See United Church of Christ Commission for Racial Justice 1987; U.S. General Accounting Office 1983.

28. Environmental justice scholars and activists also point out that the Memphis garbage strike of 1968, supported by Martin Luther King Jr., was a kind of environmental justice action (see Cole and Foster 2001). Going further back, Jane Addams's activities in Chicago in the late nineteenth and early twentieth centuries centered on poor environmental conditions and the diseases they caused (Gottlieb 1993). In addition, in 1967, African American students protested when an eight-year-old girl drowned in a garbage dump in Houston (Bullard 1994). I name the Warren County protests here because they catalyzed a concerted response from social justice activists, and environmental justice organizing grew at a rapid pace after those protests.

29. Interview with Connie Tucker, July 27, 2003. In 2002, activists held a second summit. For information, see www.ejrc.cau.edu, or www.weact.org.

30. Interview with Connie Tucker, July 27, 2003. Since this time, the EPA has reformed in many ways. In particular, Region IV has made environmental justice much more of a priority, and its relations with community members are vastly improved.

31. See Lawrence 1987; Lazarus 1993; Moss 1996. Notably, in February 2002, thirty-five hundred (mostly poor) residents of Anniston, Alabama, won a highly publicized lawsuit charging the Monsanto Company (which later became Solutia) with deliberately releasing tons of PCBs in their neighborhoods (Grunwald 2002:A1). Although this case made no race-based claims, it is a landmark in that its plaintiffs were mostly poor, and the fence-line community was predominantly African American.

32. In one recent exception, Pensacola activists won their long battle for federally sponsored relocation, marking the first occasion that a black community received permanent relocation funds. Soon after the relocation program got under way, however, community members claimed that EPA administrators assessed some residents' property values at much higher rates than others and were directing them to relocate in a predominantly low-income black neighborhood (field notes, September 1999).

33. It is tempting to refer to 1950s and 1960s black activism in the singular, as *the* civil rights movement: the activists with whom I worked often did, mainly for the sake of brevity. However, the civil rights movement encompasses multiple actions, practices, strategies, goals, and periods of time. To acknowledge that diversity, I have chosen most often to refer to a *civil rights era,* rather than a movement, which covers an approximately ten-year time span beginning in the mid-1950s, around the time of *Brown v. Board of Education* (1954) and ending in the mid-1960s, with the passage of the Civil Rights Act of 1964 and the Voting Rights Act of 1965.

34. For more information on this movement, see http://bfaa-us.org.

35. National Urban League 2004.

36. Pugh 2002:A1.

37. Gregory 1998:5.

38. Because an exhaustive review of urban poverty studies is beyond the scope of this chapter, I limit my discussion to some of the more prominent ones.

39. See Baker 1998 for a more detailed discussion of eugenics and this period in anthropological history.

40. See, for example, Boas 1938, 1945.

41. Davis, Gardner, and Gardner 1941; Drake and Cayton 1945; Myrdal 1944; Powdermaker 1939; see also Warner 1962:86–101.

42. Lewis 1966.

43. The term "underclass" was used as early as 1963 in Gunnar Myrdal's *Challenge to Affluence.* In 1977, it was popularized when *Time* magazine published "The American Underclass," an article that warned Americans about an "intractable," "socially alien," and "hostile" group of people whose "bleak environment nurtures values that are often at odds with those of the majority" (Katz 1993:4–5).

44. Hannerz 1969; see also Liebow 1967; Valentine 1968, 1969, 1971

45. Stack 1974; Ladner 1971. Throughout the 1980s, a small but committed number of anthropologists continued to offer alternative models of the urban poor that took a more nuanced view of African American life, highlighting the high levels of organization and structure required to cope with poverty (Williams 1992).

46. di Leonardo 1998:112. In addition, di Leonardo notes that capitalism creates a flexible workforce that is alternately hired and fired, depending on

corporate needs. Public policies then ensure that this workforce remains vulnerable to poverty and thus on the job market. See Harvey 1989; Piven and Cloward 1977. For other anthropological critiques of "underclass" theories, see Duneier 1998; Gregory 1998; Newman 1988, 1999; Williams 1992.

47. For concise analyses of this and other attempts to connect the "underclass" to northward migration, see Jones 1993; Trotter 1993. See also Katz 1993, whose edited volume uses historical analysis to refute underclass theories.

48. See McWhorter 2000.

49. See di Leonardo 1998:124–125; see also Duneier 1998.

50. Wilson 1978:153.

51. See Omi and Winant 1994 [1986]; Blu 1980. For instance, historians have highlighted how before slavery, phenotypic differences between people were not categorized into racial hierarchies. However, because slavery did not jibe with Revolutionary-era ideology, it became necessary to explain that people with black skin did not count as real humans and could be treated as subhuman (Roediger 1991; Jordan 1968). Early Americans also considered the Irish and Jews a separate race until they underwent a "whitening" process in the popular imagination, largely for economic reasons (see Sacks 1994). Similarly, Morsey (1994) and Takagi (1994) both show how Egyptians and Asian Americans, respectively, have been racialized, ethnicized, and reracialized according to changes in public sentiment toward minorities, as well as in their own political and economic agendas. In a more recent iteration of social constructionism, British cultural theorist Paul Gilroy (2000) argues that biotechnology and the global marketing of black culture (namely, hip-hop, rap, and the fashions they inspire) have significantly muddied the racial identity waters. Because anyone can adopt black fashions, and because genetics have been taken so far out of the realm of nature and into the realm of technology, race becomes even more of a fiction or construct.

52. Gregory 1994:22–23; Harrison 1995b:47; Sanjek 1994; see also Harrison 1998; Takagi 1994. For exceptions to this, see Mullings 1979; Ogbu 1978; and Drake 1987, all of whom continued to argue (even during the heyday of ethnicity studies) that race and ethnicity were separate categories. Also during this time, Studs Terkel began his research for *Race: How Blacks and Whites Think and Feel about the American Obsession,* which vividly illustrates Americans' notions that race is an impolite topic.

53. The solution, according to Wilson, is to create public policy that increases employment opportunities for black men so that they will be more marriageable and the nuclear family will be restored, a suggestion that Adolph Reed (1988) refers to as a "macro-economic dating service."

54. Frazier 1939; Drake and Cayton 1945. Much later, sociologist Mary Pattillo-McCoy finds that, rather than integrating white suburbs, these people

move to black suburbs located very near city limits and the places they have just come from (1999:25).

55. Wilson has also been criticized for reiterating Moynihan's sexist assumptions about the pathology of female-headed households (while ignoring questions of employment and child care for black women) and for painting a portrait of an overly idealized ghetto past that in many cases never really happened (rather, middle-income blacks who lived alongside lower-income blacks frequently exploited them). See di Leonardo 1998; Hannerz 1969; Williams 1992.

56. This information was received from the Georgia Regional Development Center in March 1999, via telephone.

57. Unless otherwise noted, Hyde Park statistics were taken from Sociology Research Methods Students et al. 1998.

58. See also Jargowsky (1997), who finds that almost half of the residents in the average high-poverty neighborhood are in fact not poor, and most poor blacks do not live in high-poverty neighborhoods. See Gregory (1998) and Newman (1988, 1999) for further discussions of the neglect of the working poor in social science.

59. Pattillo-McCoy 1999:30.

60. Sanjek 1994:1; see also West 1993.

61. Kelley 1997:172.

62. Du Bois 1935:30.

63. For further reading on the role of emotions in social movements, see Aminzade and McAdam 2001; Davies 1980; Goodwin, Jasper, and Polletta 2001; Goodwin and Jasper 2003; Jasper 1997; Morgen 2002; Reed 2004.

64. See also Lichterman 2002 on how participant observation contributes to the building of social movement theory.

65. For recent and noteworthy reviews of social movement literature, see Benford and Snow 2000; Cohen 1992; Edelman 2001; Giugni 1998; Pichardo 1997; Polletta and Jasper 2001; Strang and Soule 1998. In recent years, a number of other theoretical approaches to the study of social movements have been proposed. Political opportunity structure, for example, moves away from identities and strategies entirely, positing that "people join in social movements in response to political opportunities and then, through collective action, create new ones" (Tarrow 1994:17–18). However, this approach has been criticized for narrowly defining culture, glossing over the degree to which it penetrates activism, institutions, and social processes (Goodwin and Jasper 1999). Another group of social scientists, known as political process theorists, have answered this "culture question" through the concept of cultural framings, noting that social movement actors consciously and strategically "fashion shared understandings of the world and of themselves that legitimate and motivate collective action" (McAdam, McCarthy, and Zald 1996:6). In this view, culture comes

into play as organizers try a series of frames, or ways of presenting an issue, until they find those that resonate with potential recruits (Goodwin and Jasper 1999).

However, I also find "cultural framing" somewhat limiting. Conceptualizing social movement activities as always strategic, always cognitive, and always part of a win-lose game ignores the fact that social movements are created and enacted by people, who are guided by morals and emotions as much as they are by strategies (Jasper 1998:397).

66. For other examples of ethnographic work that combines these two approaches, see Alvarez, Dagnino, and Escobar 1998; Escobar and Alvarez 1992; Gregory 1998; Morgen 2002; Sacks 1988.

67. Edelman 2001. For RM examples, see Gerlach and Hine 1970; McAdam 1982; McCarthy and Zald 1977; Morris 1984; Piven and Cloward 1977. For critiques, see Bartholomew and Mayer 1992; McAdam, Tarrow, and Tilly 2001; Reed 1986a.

68. Dark 1998; Darnovsky, Epstein, and Flacks 1995:xv; Edelman 2001; Mahon 1997:67. For NSM examples, see Larana, Johnson, and Gusfield 1994; Melucci 1985, 1989, 1994; Touraine 1988 [1984].

69. In the past few years, many scholars have forsaken "old" versus "new" debates in favor of simply defining the various aspects of collective action (see Melucci 1998).

70. V. Taylor 1989.

71. "Strategies" comprise the tactics, strategic initiatives, and forms of political organizing that activists employed (Escobar and Alvarez 1992:4), including organized protests, proposals, lawsuits and legislative lobbying, resistance (refuting racial stereotypes), and accommodations (using multiracial language to appeal to wider audiences). "Identity" refers to an interactive and shared definition of a particular subject position that also reflects the very hierarchies of power and meaning that activists contest (Melucci 1988:342; see also Gregory 1998:11).

72. See also Lynch 1996.

73. See Edelman 2001.

NOTES TO CHAPTER 3

1. Woods, like all Masters champions, is actually an honorary member of the club.

2. Haddock 2000.

3. The city of Augusta and its neighboring Richmond County became "Augusta–Richmond County" in 1996. In this chapter, "Augusta" generally refers to the areas included in the consolidated Augusta–Richmond County.

4. Berendt 1994:31.

5. See Hyland, Register, and Gunther 1991.

6. Illustrating the "stepchild syndrome" Cashin notes that in his famous march to the sea, General William Sherman bypassed Augusta, despite the fact that it is equidistant from Atlanta and Savannah. Cashin asks, "Why did Sherman consider Atlanta and Savannah more important than Augusta?" (1980:126).

7. U.S. Bureau of the Census 2000. Between 1995 and 2000, Augusta added 4,215 new jobs, while Atlanta added 278,440 (see www.atlantaregional.com).

8. For instance, in 1999, the Georgia Department of Transportation initiated a new rail line connecting several Georgia cities to Atlanta. Augustans were dismayed to learn, however, that while Macon and Athens would receive rail lines by 2005, Augusta would not receive one until 2010 (Williams 2000).

9. U.S. Bureau of the Census 2000.

10. For my purposes, the "South" includes North and South Carolina, Virginia, Georgia, Florida, Alabama, Mississippi, Kentucky, Louisiana, Tennessee, and Florida. This region is often stereotyped as being particularly racist, and although I spend much of this book describing racial inequities, I do not mean to reinforce such stereotypes. Rather, the kinds of racial segregation and discrimination I found in Augusta are not limited to that city, its state, or its region but occur throughout the United States.

11. Hill 1998:23; see also Blu 1980; Wright 2001.

12. But see Biles's (1986) history of Memphis in the Great Depression or Lewis's (1991) history of black migration to Norfolk, Virginia.

13. See Baker 2001:20–21.

14. See Cobb 1999:61; Wright 2001.

15. Baker 2001:24; Gregory 1998:174–175.

16. Revisionist historians similarly emphasize the positive aspects of paternalism. However, as Genovese notes, "Southern paternalism, like every other paternalism, had little to do with Ole Massa's ostensible benevolence, kindness and good cheer. It grew out of the necessity to discipline and morally justify a system of exploitation" (1976:4).

17. Cashin 1980:63–64.

18. Wan 2003.

19. McCoy 1984; Wan 2003.

20. McCoy 1984.

21. Coleman 1978:97.

22. Cashin 2001:36.

23. Coleman 1978:85, 95.

24. Moore 1905.

25. McCoy 1984.

26. See Silver and Moeser 1995 for a historical account of the relationship between segregation and black self-sufficiency and activism in key southern cities.

27. McCoy 1984:30–33; Wan 2003.

28. A total of 110 graduates of one black private school alone enlisted in the army (McCoy 1984:35).

29. Goldfield 1987; Schulman 1991.

30. Davis, Gardner, and Gardner 1941; Myrdal 1944; Powdermaker 1939.

31. Lewan and Barclay 2001. More recently, in 1997, the U.S. Department of Agriculture (USDA) Civil Rights Action Team (CRAT) revealed that these activities continued well into the 1980s and early 1990s. After settling a 1999 lawsuit, the USDA has paid more than $630 million to farmers and former farmers who could document that they were unfairly denied loans (see http://www.oxfamamerica.org/advocacy/art4066.html).

32. McMillen 1990:149; see also Conrad 1965; Davis, Gardner and Gardner 1941:337; Powdermaker 1939; Schulman 1991.

33. Rowland and Callahan 1976.

34. Brown with Tucker 1996:9.

35. Biles 1994:37, 104; Schulman 1991:20.

36. As a result of this and other kinds of discrimination, the NRA became known as the "Negro Removal Act" and "Negroes Ruined Again." For examples of the extensive literature on African Americans and the New Deal, see Biles 1990, 1994; Bunche 1973; John Kirby 1980; Jack Kirby 1987; Sitkoff 1978; Weiss 1983; Wright 1986; Wolters 1970.

37. U.S. Bureau of the Census 1940. Toward the end of the 1930s, Eleanor Roosevelt and other New Deal liberals stepped in and called for more attention to the conditions of blacks. Yet, while they were able to improve housing and employment conditions in the North, the South stayed entrenched in the brutality of Jim Crow laws and practices.

38. Biles 1994:104. The literature on the "Great Migration" of African Americans northward is vast. For some cogent examples, see Grossman 1989; Lemann 1991; Henri 1975; Trotter 1991.

39. Brown with Tucker 1996:9. See also McCoy 1984:62.

40. Cobb 1975:98.

41. Brown with Tucker 1996:14.

42. Rowland and Callahan 1976; see also Lutz 2002.

43. Cobb 1975:98; see also Cashin 1980:297; Cobb 1982; McCoy 1984.

44. Biles 1990:95.

45. Chamber of Commerce, Augusta, Georgia, 1947.

46. Like any version of the "good old days" Striggles's may paint his portrait of 1950s life with a rose-tinted brush (see di Leonardo 1998; Hannerz 1969; Williams 1992).

47. For a well-informed and detailed analysis of the effects of highway construction on racial segregation in Atlanta, see Bayor 1988. For additional analyses of the effect of new highways on the lives of poor and minority urban residents, see Hirsch and Mohl 1993; Jackson 1993; Sugrue 1993.

48. See Dudziak (2000), who points to McCarthyism as a major factor in stimulating 1960s protests. McAdam (1982), on the other hand, takes a different tack and argues that as the United States and the Soviet Union competed for the allegiance of third world countries, the United States recognized the "propaganda value of resolving racism" and passed the Civil Rights Act of 1964 and the Voting Rights Act of 1965. Other scholars argue that segregation constricted blacks' consumer power and impeded the goals of monopoly capitalism (Reed 1986a:71). Still others contend that southern African Americans were politicized by their experiences in World War II and the Korean War. In addition, post–World War II economic prosperity and the creation of new employment opportunities in urban centers stepped up rural-to-urban migration trends. Growing concentrations of blacks in cities increased membership in urban black churches and attendance at historically black colleges. African Americans thus found new opportunities for forming coalitions, which then swelled into the protests of the civil rights era (McAdam 1982; McDonogh 1993).

49. Payne 1995.

50. On February 1, 1960, four black North Carolina teenagers staged a sit-in at a Woolworth's lunch counter. This event sparked a sit-in movement throughout the South between February and May of that year.

51. This time their efforts led to a lawsuit, and in 1962 a federal judge ordered an end to segregation on city buses (Wan 2003).

52. Cashin 1980:110–111, 298–300.

53. Cashin 1980:299; 1985. Later, opposition to the Civil Rights Act of 1964 was so intense that enough white Democrats switched party allegiance to give Barry Goldwater a majority of Georgia's votes in the 1964 presidential election. Georgians also elected the first Republican congressman since Reconstruction (Coleman 1978:103).

54. Cashin 1980:298.

55. Richmond County school superintendent, Roy Rollins, and the school board president, W. R. Loflin, also publicly maintained that segregation must continue "at any cost" (Cashin 1985:111).

56. Bayor 1996.

57. Wan 2003.

58. Cashin 1980:303.

59. See Wright 1986.

60. See Lipsitz 1995 [1988]. Similarly, participation in World War I led some African Americans to create the Niagara Movement and, later, the National Association for the Advancement of Colored People (Ortiz 2000). Participation in World War II led to the "Double-V" (victory over the Nazis and victory back home) campaign in the 1940s.

61. Holton 1969:A1.

62. *Augusta Chronicle,* May 12, 1970; Wan 2003.

63. Smoller 1970:A1.

64. Cashin 1980:302.

65. Alston 1970. The commission investigates cases of racism brought before it by local citizens, but it has no power to enforce any of its findings.

66. There are several "New South" periods. Another well-known one occurred just after the Civil War and was similarly marked by an influx of northern industries into to the South.

67. Wright 1986:259.

68. Wright 1986; see also Bullard 1989.

69. Cogan 1982; see also Bullard 1989.

70. Bullard 1989; see also Stack 1996.

71. Rural Development Center 1975:182; 1985:104; 1999:143.

72. See also Lewis (1991), who describes wide variations in migration patterns to the North.

73. Rural Development Center 1999:43, 47.

74. Rural Development Center 1999:35.

75. Katz 1993:454; see also Jones 1993.

76. Interview with Walt Alexanderson, Thermal Ceramics public relations director; 2004 figure courtesy of the Augusta Chamber of Commerce.

77. Regional Development Center in Augusta, Georgia, via telephone interview, March 1999.

78. Cooper 1998:6A.

79. Pavey and Johnson 1998.

80. For instance, in the late 1980s, federal officials found that the Atlanta Gas Light Company plant, located in an African American neighborhood in downtown Augusta, was leaking coal tar. In 1997, in response to a lawsuit filed by the Trinity Christian Methodist Episcopal Church (a historic African American church) and two homeowners, Atlanta Gas Light launched a $50 million cleanup effort (Pavey 2001). Because the case was fairly clear-cut, Atlanta Gas Light Company settled out of court.

81. To better visualize Hyde Park, it is helpful to imagine what anthropologist Rhoda Halperin refers to as a "shallow urban" area. A shallow urban neighborhood is spread out, not overcrowded, easily accessible to the city but separate and separated from it. Geographically and culturally, it lies somewhere between the crowded, pavement-laden city and "shallow rural" areas, where small farms and homesteads with gardens line interstate highways that lead to factories, flea markets, malls, and eventually the city.

82. This figure is based on my own door-to-door survey of Hyde Park houses in October 1999.

83. In 1999, a homeless man, whom no one claimed to know, was found dead in a house on Walnut Street.

84. See Edelstein (2003) on coping with ongoing toxic danger.

85. Sociology Research Methods Students et al. 1998:14.

86. See also Halperin (1998), who offers a detailed analysis of such resource sharing, known as "householding."

87. Established in 1954, Christian Fountain originally drew most of its members from the neighborhood. In 1998, however, about half of Christian Fountain's members were Hyde Park residents, and half came from other areas. This diversity of membership is typical of other local southern black Baptist churches, which have also increasingly lost local members while gaining worshipers from other neighborhoods. In Christian Fountain's case, residents fondly recall the church's original pastor as a dynamic man, heavily involved in the community. After he died, congregants had difficulty finding and keeping replacements, and many found other places to worship.

NOTES TO CHAPTER 4

1. Paolisso and Maloney 2000.

2. Johnson 1994:11.

3. See Basso 1996; Halperin 1998; Lynch 1994:49.

4. Johnson 1994:39; Reidy 1993; Westmacott 1992; see also Berlin and Morgan 1993.

5. Westmacott 1992; see also Genovese 1976. For further descriptions of slave and sharecropper gardens, see Johnson 1941; LeMaistre 1988; Powdermaker 1939; Raper 1936.

6. It is also possible that gardens, and the produce they generated, represented the warmth and nourishment that food typically signifies, and that the act of eating garden produce together with kin and neighbors symbolized community. However, as I mentioned, Hyde Park residents stressed the links that gardens had to the American Dream.

7. Stack 1996:42.

8. Stack 1996:xv–xvii.

9. Gregory 1998:55; see also Massey and Denton 1993:479.

10. Perin 1977; see also Edelstein 2003:62; Gregory 1998; Harrison 1995a: 39; Hochschild 1996:42, 47. As Perin also notes, President Franklin D. Roosevelt once commented that a nation of homeowning families is "unconquerable." Similarly, Presidents Calvin Coolidge and Herbert Hoover both claimed that homeownership was our nation's greatest ideal.

11. See Huth 1991a, 1991b. For more information on these tests and their results, see chapter 5. Here it should be noted that the UGA studies were highly disputed.

12. Deacon Saulsberry claimed that if rats continued to live in Hyde Park without getting sick, then humans were probably also safe. Moreover, he argued that as long as he cooked his vegetables at high temperatures, any chemicals

they may contain would be killed. A few other elderly residents who did not believe in the contamination said they had let their gardens die because they had gotten too old to tend them, but these people were certainly exceptions.

13. Prior to the Clean Water Act of 1977, industries typically dumped their wastewater into unsecured pools, which often leaked into nearby areas.

14. SWP did not sanction such activities. Furthermore the ATSDR has stated that none of these activities could cause adverse health effects. See ATSDR 1995.

15. Halperin 1998:310. For further examples of how external surroundings are incorporated into community identities, see Daniel 1984; Lynch 1996.

16. See Edelstein (2003), who finds a similar ambivalence over home investment in his study of the effects of contamination on the Legler section of Jackson township, New Jersey.

17. Edelstein 2003:65.

18. In 1990, environmental issues came close to home when the Army Corps of Engineers halted a drainage project just outside of the Hyde Park area because it endangered wetlands. Interestingly, HAPIC activists took no part in advocating either to continue the drainage project (which may have helped reduce flooding in the neighborhood) or to stop and preserve the wetlands.

19. Bullard 2000:28, 31.

20. See Agarwal 1992; Enloe 1989; Krasniewicz 1992; Hamilton, 1990; Harvey 1996; Simpson 2002; Szasz 1994; Zeff, Love, and Stults 1989. See also Checker 2004b for a more in-depth discussion of my own findings regarding women and environmental justice activism.

21. For example, the Clean Water Act and the reauthorization of the Federal Insecticide, Fungicide, and Rodenticide Act were passed in 1972, and the Resource Conservation and Recovery Act was passed in 1976.

22. Most of the information in this paragraph was taken from the "Southern Wood Piedmont Augusta Plant History," a document prepared by SWP's attorney.

23. However, these illnesses were never proved to correlate with the toxic chemicals found in residents' groundwater.

24. Governor's Task Force Report 1996.

25. SWP disputes this claim, stating that it discontinued the use of creosote in 1983, and that proper sampling techniques were not used.

26. Hewell 1989:8A.

27. Hewell 1989:1A.

28. Virginia Subdivision residents made similar allegations, and some said that they had photographed such activities; however, I never saw any such photographs or other documentation of illegal dumping, and SWP vehemently denies that it ever took place.

29. SWP's attorney points out that the company instituted an extensive remediation program, constructing a slurry wall around the site that is impervious

to water, excavating the most contaminated parts of the soil, and flushing hundreds of thousands of gallons of groundwater into the county's sewage system.

30. I borrow this subtitle from "Erin Brockovich Doesn't Live Here," an article by anthropologist Mark Moberg (2002). *Erin Brockovich* is a film that appeared in 2000 about a white working-class town that was contaminated by big industry and then successfully saved via a lawsuit initiated by a paralegal in a nearby law firm. Moberg's point, and mine, is that such "Hollywood endings" rarely happen in African American neighborhoods.

31. The federal judge overseeing the case was also a shareholder in Merry Land and Development Company (one of the plaintiffs in the original lawsuit filed by Virginia Subdivision and surrounding property owners). Because of that, he recused himself from that case. However, he did not believe that he had a conflict of interest in the Hyde Park case. Nonetheless, Hyde Park's attorneys filed an unsuccessful motion to have him recused from their suit.

32. Ollie Jones passed away from a heart attack in 2001 at the age of fifty-five.

33. See Gwaltney 1993; Prince 2002; White 1999; Williams 1974, who document similar findings about viewing certain issues through a race-identified lens.

34. See, for example, Anderton et al. 1994.

35. Massey and Denton 1993.

36. Cole and Foster 2001; see also Bullard 2000.

37. Sociology Research Methods Students et al. 1998.

38. HAPIC leaders reported that at one point, they had asked the Augusta Sierra Club for help but received only "words of encouragement."

39. In terms of drug traffic, Hyde Park's dealers and their customers did not reflect national trends. Whereas in the late 1990s, across the United States, marijuana had replaced crack as the drug of choice sold in inner-city urban areas, in Hyde Park crack replaced heroin in the 1980s and remained popular throughout the 1990s.

40. To be fair, I did not live in the neighborhood during my fieldwork, and residents often told me that after dark everything changed and it became a much rougher place. Although few crimes occurred during my fourteen months of fieldwork, it is possible that they went unreported, although it would be very surprising not to have heard about such incidents through gossip. My findings here diverge from those of Pattillo-McCoy (1999) who also studied close-knit community relations in an African American urban neighborhood marked by criminal activity. However, Pattillo-McCoy found that because they had grown up with drug dealers and gang members, most community residents felt safe and even protected from potential violent crimes. Importantly, in Pattillo-McCoy's neighborhood, drug dealers were more powerful than in Hyde Park, as they occupied positions far higher on the local hierarchy of drug-selling networks.

41. Sociology Research Methods Students et al. 1998:20.

42. Not all residents held such idealized views. Some admitted that while they would not mind being relocated with certain of their neighbors, they would rather others moved elsewhere.

43. "Mother Utley" refers to Mary Lou Utley. Reverend Roundtree was a pastor at Christian Fountain. During the early 1990s he became very active in HAPIC efforts.

44. Edelstein 2003.

45. Similarly, historians tell us that as they fought for suffrage in the late 1800s, African American activists developed expansive agendas that included equal pay, education, and citizenship rights (Montgomery 1967).

46. See also Harvey 1996:263.

NOTES TO CHAPTER 5

1. The plaque was a gift from Charles Utley.

2. The subheadings in this chapter are taken from HAPIC flyers and newsletters, which always contained messages to inspire and encourage community members.

3. For example, a 2001 video made by an Augusta State University student on black women leaders in the Richmond County region was dedicated to Mary Utley for her work in Hyde Park.

4. See Besson 1995.

5. Light 1972.

6. Cf. Gregory 1998; Scott 1990:5; Stack 1996.

7. See also Reed (1999:120), who describes how in the 1960s, struggles over such infrastructure were "the stuff of intense struggle and confrontation" in black communities.

8. See also Piven and Cloward (1977), who argue that as the post-Reconstruction years wore on, African Americans increasingly compared their lives with those of whites, and these comparisons fed the activist explosions of the 1950s and 1960s.

9. See Gerlach and Hine 1970; McAdam 1982.

10. Like many community groups of this period, the name of this organization bore a close resemblance to Martin Luther King Jr.'s Montgomery Improvement Association (Reed 1999).

11. Lipsitz 1995 [1988].

12. Political theorist Adolph Reed (1999:19) connects these discourses to the New Deal, which generated an egalitarian ideal later expressed as demands for full citizenship.

13. Gerlach and Hine 1970; see also McAdam 1982. Conversely, see Adolph Reed (1986a, 1999), who contends that these structural approaches do not pay enough attention to dynamics and agency within black communities.

14. See Bartholomew and Mayer 1992; Goldberg 1991; Gregory 1998; Lipsitz 1995 [1988]; Piven and Cloward 1977.

15. In 1998–1999, Jenkins (a K–5 school) enrolled only approximately ninety total students. In the fall of 2001, the school closed and became a Head Start program.

16. "Resource mobilization" theorists, in particular, have been accused of this. See Buechler 1995; Cohen 1992; Edelman 2001; Escobar and Alvarez 1992; V. Taylor 1989.

17. See V. Taylor 1989; Rupp and Taylor 1987.

18. *Jordan et al. v. Southern Wood, ITT Rayonier, and ITT Corp.*, a.k.a. "Jordan et al. v. Southern Wood et al."

19. Cole and Foster 2001: 63.

20. For a more detailed analysis of this case, see Cole and Foster 2001:64, 70.

21. Title VI states, "No person in the United States shall, on the ground of race, color, or national origin, be excluded from participation in, be denied the benefits of, or be subjected to discrimination under any program or activity receiving Federal financial assistance" (cited in Lazarus 1993:834; see also Colopy 1994).

22. For other critiques of NEPA and Title VI, see Bullard 2000; Cole and Foster 2001; Lazarus 1993.

23. Tesh 2000; Wigley and Schrader-Frechette 1996; Zimmerman 1994.

24. Israel 1995:486; Schettler, Barrett, and Raffensperger 2002; see also Douglas and Wildavsky 1982.

25. Agency for Toxic Substances and Disease Registry 1994.

26. West et al. 1992.

27. See Fitchen 1988; Montague 2003; Novotny 1998:141.

28. Wigley and Schrader-Frechette 1996; see also Anglin 1998; Bryant 1995b; Novotny 1998.

29. U.S. Environmental Protection Agency, Risk Assessment Forum 1986.

30. Israel 1995:506. For an excellent (and frightening) example of how science fails to measure adequately human exposure to toxins, see Schrader-Frechette's (1999) analysis of how environmental scientists evaluated the effects of Chernobyl.

31. Anthropologist Emily Martin (1994) argues that we mistakenly tend to see scientific knowledge as flowing exclusively in one direction—from scientists to passive recipients within a culture. Instead, we might recognize that scientists both inform *and are informed by* the culture of which they are a part. For examples of critical science literature that is particularly useful for environmental justice, see Bryant 1995b; Franklin 1995; Haraway 1989, 1991; Jasanoff et al. 1995; Nelkin 1987, 1992; Satterfield 1997.

32. Cutter, Scott, and Hill 2002:420.

33. O'Brien 2000.

34. The reasons for this are contested. Virginia Subdivision residents claim that after paying off legal fees, they did not have enough money left over to move; but SWP spokespersons maintain that settlement amounts were distributed according whether a particular property was found to be contaminated by wood-treating chemicals. Some African American residents of Virginia Subdivision also claimed that they had not received as much compensation as white families.

35. I further discuss diversity organizing in the following chapter.

36. Cardoso 1992:292; see also Lynch 1996.

37. *Metro Courier* 1991:1A.

38. See Craig 1991a:1D. The Georgia EPD launched a criminal investigation into the matter that eventually found no evidence of SWP's wrongdoing (Donahue 1992c:1A).

39. ATSDR 1994.

40. Craig 1991c:1D. SWP disputes the validity of these studies, on the grounds that they did not follow proper EPA protocols. But, as I mentioned earlier, those protocols are quite expensive. Less costly alternatives may indeed be just as valid as EPA protocols, but in legal arenas, without the EPA's official blessing, such studies are easily disputed.

41. Anglin 1998; Novotny 1998; Satterfield 1997; Tesh 2000.

42. Sociology Research Methods Students et al. 1998.

43. See Latour's 2004 essay on the pitfalls of postmodern critique, especially in the context of contemporary leftist politics.

44. Donahue 1992a.

45. Governor's Task Force 1996.

46. This quotation was taken from the documentary video *Long Is the Struggle, Hard Is the Fight: The Hyde Park Story,* produced by Project South, an Atlanta-based nonprofit specializing in community organizing.

47. See Cole and Foster 2001.

48. Donahue 1992b.

49. This is a skin condition, resulting from long-term arsenic ingestion. Affected individuals develop multiple lesions, commonly on the palms and soles but also on the fingers and proximal portions of the extremities.

50. I was unable to secure copies of either the dermatology or the neurological study; I report their results as summarized in Governor's Task Force Report 1996.

51. Donahue 1992b.

52. For alternative research models that emphasize community knowledge and control, see Brown 1992; Bryant 1995b; Kroll-Smith and Floyd 1997; Wisner 1997.

53. Pavey 1993:1A; see also Langford 1993.

54. ATSDR 1994.

55. Interview with John Jenkins, July 9, 1999.

56. Activists cited many reasons for this belief, ranging from Clinton's being raised by a single mother in the South to his skill at the saxophone. Many also believed that like several black elected officials (e.g., Marion Barry), Clinton was set up by a white Republican power structure during the "Lewinsky scandal."

57. Other soil and groundwater samples had been taken in previous years, but my focus in this section is on the EPA study. Further details of the results summarized here can be found at www.atsdr.cdc.gov/HAC/PHA/swp/swp_p4c.html.

58. ATSDR 1994. This practice, known as "geophagy," or "pica," does occur in places around the world, for a variety of reasons.

59. I was unable to obtain further information about the specific nature of this evidence.

60. The boycott never went very far, since pinpointing ITT's targetable subsidiaries proved difficult, and most grassroots activists could not afford to stay in Sheraton Hotels anyway.

61. Unfortunately, after the protest, church support dwindled.

62. Bush Field's security officers initially refused to let picketers onto the airport grounds. After attorney Bill McCracken threatened to charge the mayor's office with restricting freedom of speech, however, the picketers were allowed to protest.

63. Pavey 1994:11A.

64. Pavey 1994:11A.

65. Melucci 1988:248.

66. Escobar 1992:73. This perspective contrasts with that of traditional "resource mobilization" theorists such as Gerlach and Hine (1970), McCarthy and Zald (1977), and Piven and Cloward (1977), who evaluate outcomes of social movement activities in objective terms rather than in terms of how activists themselves define their successes.

67. During the latter part of my fieldwork, this grant project, known as the Health Intervention Project, was under way. Preliminary results did not reveal significant health issues, but health officials have decided to apply for more funding to conduct further study.

68. See, for example, Lipsitz's (1995 [1988]) in-depth profile of Ivory Perry, a civil rights activist who eventually ran for elected office.

69. Pavey 1998:1C.

70. All these leaders were male. Although HAPIC's leadership had traditionally been equally divided among men and women, due to personal conflicts, its female leaders became less active in the late 1990s. Even though neither side of the conflict described it as a gender issue, gender is an important topic in the study of environmental justice. For my own analysis, see Checker 2004b.

71. Office of Solid Waste and Emergency Response 2002.

72. More recently, thanks to the efforts of grassroots environmental justice activists, the Brownfields Initiative was amended to include residential areas.

73. Sociology Research Methods Students et al. 1998.

74. Gregory 1998:98.

75. According to Gramsci (1991), social groups create different strata of intellectuals who emerge from the group and act as orators, organizers, and active participants in practical life. These people also have different degrees of connection back to the social group. In the case of HAPIC, certain activists who were particularly facile at establishing connections on various extralocal levels emerged as such "organic intellectuals" and led the organization's activities in the late 1990s. My focus here, however, is on they ways in which these positions are conflicted within a community group. For a similar discussion of community leaders as "insider-outsiders," see Halperin 1998:293.

76. Elsewhere I argue that class is not the transparent category we take it for, and its definition shifts according to context. Observable criteria—income, wealth, occupation, and education—work in conjunction with less obvious criteria such as a person's behavior, perspective, and accountability. These criteria are then judged according to specific cultural values; see Checker 2004b. For other recent anthropological work on the social construction of class categories among African Americans, see Gregory 1998; Jackson 2001.

77. Reed 1986a:71; see also Gaines 1996; Gatewood 1990.

78. Checker 2004b.

79. Bullard 2000; Cole and Foster 2001; Harvey 1996; Pena 2002. See also McAdam, Tarrow, and Tilly 2001; Tarrow 1998 on "framing," or the discursive practices that shape social movement actors' understandings of, and opposition to, their conditions.

80. Alvarez, Dagnino, and Escobar 1998; Escobar and Alvarez 1992:8; Polletta 2002; Rupp and Taylor 1987.

81. See also Escobar and Alvarez 1992; Gregory 1998; Lynch 1996.

82. Melucci 1994; see also Burdick 1998.

83. Sociologist Verta Taylor calls organizations like HAPIC "abeyance structures," or institutions through which a group "can maintain its identity, ideals and political vision" (1989:772; see also Rupp and Taylor 1987). By providing a space for movements to hibernate, retrench, and adapt to changing political climates, such structures become important resources for subsequent mobilizations.

84. Polletta 2002; see also Reed, who writes, "Democratic and participatory values must be the cornerstone of credibility for the notion of black politics; group consensus must be constructed through active participation" (1999:49).

NOTES TO CHAPTER 6

1. See also Checker 2004a, which contains some of the same information and arguments presented in this chapter.

2. See Alley, Faupel, and Bailey 1995; Cole and Foster 2001; Pena and Mondragon-Valdez 1998; Pulido 1998; Stonich and Bailey 2000. For some examples of potential exceptions to this, see Pena and Mondragon-Valdez 1998; Stonich and Bailey 2000. For further studies of grassroots-professional environmental alliances, see Bebbington 1993; Haenn forthcoming; Moberg 2001; Paolisso and Maloney 2000; Pulido 1996; Rangan 1993; see also Satterfield 2002.

3. For instance, the files and folders we see on our "desktops" are particularly Western conventions, as are the premises that underlie basic computer codes (Warschauer 2003). In addition, various hierarchies of connections (such as Internet service providers, which connect to larger Internet routing points, which in turn connect to even larger routing points) are controlled by a handful of elite corporations (Ribeiro 1998).

4. Quoted in Scheffer 1991:19.

5. Quoted in Darnovsky 1992:38; see also Tarlock 1994:483.

6. The rest of the Big Ten are the Audubon Society, the Natural Resources Defense Council, Friends of the Earth, the Wilderness Society, the Environmental Policy Institute, the Izaak Walton League, and the National Parks and Conservation Association. For further analysis of this "wave" of environmentalism, see Cole and Foster 2001; Gottlieb 1993; Pulido 1996; Schwab 1994.

7. Gottlieb and Ingram 1988:122.

8. A recent iteration of such movements occurred in 1998 when the Sierra Club initiated a proposal to lobby the U.S. government to limit immigration. Club members later defeated the proposal in a national election.

9. I place this peak between 1968 and 1972, around the time of the first Earth Day and a period when movement goals shifted from conservation to more holistic concerns such as pollution, nuclear power, and so forth (Tarlock 1994:480).

10. See Bryner 2001; Bullard and Wright 1987:180; Gottlieb 1993.

11. Bullard 2000:4.

12. Bullard 1993:12. Very public altercations over logging and forestry have also called attention to divisions between mainstream environmentalists and blue-collar workers; see Satterfield 2002.

13. Bullard 1993:11; see also Austin and Schill 1994:70.

14. Minority Opportunities Study as cited in Schwab 1994; see also Rothman 1988.

15. Hornblower 1997:67. Following up on this new emphasis, in 1993, the national Sierra Club hired an environmental justice program coordinator, and in 2000 it assigned four local environmental justice organizers in Detroit, Memphis, Los Angeles, and Washington, D.C. A few years later, it added central Appalachia and Arizona to that list.

16. See also Checker 2001.

17. Appadurai 1996; Goldschmidt 2004; Harvey 1989; Jameson 1991; Lowe 1996; Turner 1993.

18. However, a number of social scientists caution that the use of purely celebratory multiracial or multicultural rhetoric often masks the continued existence of racial hierarchies. See Balibar 1991; Gregory 1994; Harrison 1995b:49; McLaren 1994; Takaki 2002; Shankar 2004.

19. Cross-cultural approaches to computers are a relatively new area of anthropological research. Those who have studied the issue have found that responses to technology occur within "the context of deeply-held culturally supplied narratives" (Pfaffenberger 1995:78; see also Ingold 1997; Miller 2000; Strathern 1997).

20. *Philanthropy News Digest* 1999. But most of these donations funded programs aimed at children, despite the fact that children already were using computers in school.

21. The Savannah River Site was a DOE-owned nuclear weapons facility located twenty miles from Augusta. I provide more information on HAPIC's involvement with the site in the penultimate section of this chapter.

22. See also Warschauer 2003. Importantly, many other anthropologists studying international social movements have noted the ways in which Internet access is used by activists and facilitates their organizing; see Castells 1997; Miller 2000; Nash 1997.

23. Cooper 1993:15A.

24. I base this statement on my own conversations with environmental justice activists from different parts of the South. Similar findings can be found in Bullard 2000; Cole and Foster 2001; Liebow 1988; Novotny 1998; Roberts and Toffolon-Weiss 2001; Wolfe 1988.

25. Stoffle et al. 1988.

26. Few studies have specifically addressed the connections between religion and grassroots environmental justice activism in the United States. For some recent examples, see Arp and Boeckelman (1997), who find that church attendance was highly correlated with environmentalism for African Americans in Louisiana; see also Bullard 2000; Townsend 2000.

27. See also Regan and Legerton 1990; Whiteley and Masayesva 1998. Overall, the connections between religion, spirituality, and the environment form the basis of a relatively recent subgenre of environmental anthropology, known as spiritual ecology. See http://www2.soc.hawaii.edu/css/anth/projects/thailand/spiritualecology.htm.

28. Genovese 1976; see also Frazier 1963; Morris 1984; Myrdal 1944. For some first-person narratives on the topic, see Johnson Reagon 1993; Williams 1974.

29. Kelley 1994:43. More specifically, in his account of black union organizing during the 1930s, Kelley finds that black workers integrated religion and

organizing, and often these workers would "turn union gatherings into revival meetings" (41).

30. Lincoln and Mamiya 1990; McAdam 1982.

31. Reed 1999:61; see also C. Carson 1995 [1988]; Marx 1979. Historian Manning Marable (1983) takes a more moderate approach, arguing that the role of the church in black politics is highly ambiguous and walks a fine line between accommodation and change.

32. Morris 1984.

33. Gregory 1998:146.

34. Baer and Jones 1992; Hall and Stack 1982; Stack 1996. See also Checker 2004a for a more detailed analysis of African-based spiritualities and environmental justice activism.

35. Gottlieb 1993.

36. Haddock 1999:3.

37. Schill 1998.

38. Cline 1998.

39. Here Dicks refers to President Clinton's 1997 apology for government-sponsored experiments at the Tuskegee Institute, where African American men with syphilis were studied without being treated.

40. It was common for African American activists throughout Augusta to refer to men in positions of power and respect as "doctor."

41. Most of the activists in attendance were men, and only male HAPIC activists asked questions. During my fieldwork, HAPIC leadership was male dominated, and this stood out as another contrast with the leadership of the professional groups with which it worked. For a discussion of gender in the environmental justice movement, see Checker 2004b.

42. The MOX process was eventually approved for SRS.

43. Ribeiro 1998:345; see also Escobar 1995.

44. Alvarez (1998) for instance, finds that in the international women's movement nongovernmental organizations (NGOs) attempted to fit grassroots activists into certain middle-class feminist molds that did not necessarily accord with grassroots groups' own interpretations of feminism. For additional critiques of NGOs, see Escobar 1995; Fisher 1997; Gezon 2000; Markowitz 2001; Mayo and Craig 1995; Rahman 1995; Roberts 2000.

45. Polletta 2002.

NOTES TO CHAPTER 7

1. This material also included 928 tons of hazardous grinding waste; 13,325 tons of excavated nonhazardous contaminated soil; 807 tons of tires; 1,444 tons of excavated metal; 296 tons of petroleum-contaminated soil; 4,462 gallons of contaminated surface water; 1,000 empty compressed gas cylinders; 2,333 tons

of excavated debris; and 101 tons of PCB and metal-contaminated hazardous waste (see http://www.augustaga.gov/departments/mayors_office/media/200000 EPA.pdf). Activists and one of HAPIC's attorneys claimed they had evidence that Goldberg had made a deal to store some of SWP's waste; however, I have never seen this evidence.

2. This information is also taken from the 2001 Augusta Brownfields Redevelopment Project supplemental grant application. The chemicals mentioned were abated by excavating the soil containing them and replacing it with new soil. According to investigative reports, the contaminants remained fairly close to the soil surface.

3. Residents' belief that they were invisible to the rest of Augusta was borne out by the fact that the neighborhood was not marked on any map. For instance, during my fieldwork, one of HAPIC's major projects was to create and install signs announcing to people that they were entering the neighborhood. Moreover, when I first came to Augusta in 1998, I stopped in at the local tourist information office and asked for directions to the neighborhood, only to find that the staff person on duty had never heard of Hyde Park.

4. Here I refer to W. E. B. Du Bois's famous quote describing the last chapter of *The Souls of Black Folk* as "a tale twice told and seldom written" (1896 [1903]:5). For my purposes, the "tale twice told" highlights the fact that despite the long history of black activism, and the legislative gains made in the 1960s, civil rights battles continue to be waged every day throughout the country.

5. National Urban League 2004.

6. See also Du Bois (1896 [1903]) and Baker (2001), who make cogent arguments for why studying and writing about African American life explains American life.

7. Here I borrow from Ulrich Beck's famous comment that "smog is democratic" (see Beck 1992 [1986]).

NOTES TO APPENDIX A

1. See www.hapic.org.

2. This tradition has roots as far back as the nineteenth-century ethnologist Frank Cushing, who in studying a Zuni pueblo notes the value of reciprocal fieldwork (see Cushing 1979). Franz Boas, Margaret Mead, Henry Schoolcraft, Hortense Powdermaker, and countless other American anthropologists also practiced reciprocal and activist ethnography. For some examples of works that review and discuss anthropology's "hidden activist agenda," see Eddy and Partridge 1978; Foley 1999; Ginsburg and Tsing 1990; Marcus and Fisher 1986. For some (though certainly not all) more recent discussions of "engaged" anthropology, see Checker and Fishman 2004; Lassiter 2003; Singer 2000; Young 2001.

3. Rosaldo 1993:21. See also Marcus and Fisher's *Anthropology as Cultural Critique* (1986), a seminal work that launched a critique of the methods, rhetoric, and paradigms of anthropological study.

4. I did not live in Hyde Park. At the time, HAPIC activists claimed that there was no suitable housing for me (most vacant houses were in significant states of disrepair) and that they did not want the presence of toxic chemicals to endanger my health. I chose to honor these requests and lived approximately six miles from the neighborhood, visiting it daily.

5. My presence in Hyde Park sparked a similar skepticism among certain people outside the neighborhood as well. More specifically, upon one of my return visits, I was leaving Hyde Park with a friend in her sports utility vehicle. We were pulled over by a policeman who wanted to know where we had been in the neighborhood and why we had been there.

6. Many residents did not own cars and, due to Augusta's unreliable bus system, relied on neighbors and friends to take them places.

7. For example, during my fieldwork, the Monica Lewinsky scandal befell the Clinton White House and became the subject of much discussion in Hyde Park, as in the rest of the country. Residents' attitudes toward the scandal showed many similarities and had much to do with their racial identities. Similarly, residents often discussed the politics of family planning, as well as recent welfare reforms, issues that also provided me with rich ethnographic information.

8. Name changing is becoming a controversial and intriguing topic in anthropology, especially among those of us who study activists. For example, in her study of the women's health movement, Sandra Morgen (2002) similarly uses a combination of real names and pseudonyms. In his study of activists in Queens, Steven Gregory (1998) uses mainly real names.

9. Sociology Research Methods Students et al. 1998:2.

Bibliography

Adamson, Joni, Mei Mei Evans, and Rachel Stein, eds. 2002. *The Environmental Justice Reader: Politics, Poetics, and Pedagogy.* Tucson: University of Arizona Press.

Agarwal, Bina. 1992. "The Gender and Environment Debate." *Feminist Studies* 18:119–158.

Agency for Toxic Substances and Diseases Registry (ATSDR). 1994. "Health Consultation Final Release."

Augusta–Richmond County, Ga.: U.S. Department of Health and Human Services, March 3.

———. 1995. "Petitioned Public Health Assessment Addendum." Augusta–Richmond County, Ga.: U.S. Department of Health and Human Services, November 21.

Alley, Kelly D. 2002. *On the Banks of the Ganga: When Wastewater Meets a Sacred River.* Ann Arbor: University of Michigan Press.

Alley, Kelly, Charles Faupel, and Conner Bailey. 1995. "The Historical Transformation of a Grassroots Environmental Group." *Human Organization* 54: 410–416.

Alston, John. 1970. "Black Leaders, City Mold 6-Point Accord." *Augusta Chronicle,* May 14, A:1.

Alvarez, Sonia. 1998. "Latin American Feminisms Go Global." In *Cultures of Politics/Politics of Cultures: Re-visioning Latin American Social Movements.* Sonia Alvarez, Evelina Dagnino, and Arturo Escobar, eds., Pp. 293–324. Boulder, Colo.: Westview.

Alvarez, Sonia, Evelina Dagnino, and Arturo Escobar, eds. 1998. *Cultures of Politics/Politics of Cultures: Re-visioning Latin American Social Movements.* Boulder, Colo.: Westview.

American Lung Association. 2002. "Asthma: An Impact Assessment." September.

———. 2004. *Trends in Asthma Morbidity and Mortality.* April. New York: Epidemiology and Statistics Unit Research and Scientific Affairs.

Aminzade, Ron, and Doug McAdam. 2001. "Emotions and Contentious Politics." In *Silence and Voice in the Study of Contentious Politics.* Ronald R.

Aminzade, Jack A. Goldstone, Doug McAdam, Elizabeth J. Perry, William H. Sewell, Jr., Sidney Tarrow, and Charles Tilley, eds. Pp. 14–50. Cambridge: Cambridge University Press.

Anderton, Douglas L., Andy B. Anderson, John Michael Oakes, and Michael R. Fraser. 1994. "Environmental Equity—The Demographics of Dumping." *Demography* 31:229–248.

Anglin, Mary. 1998. "Dismantling the Master's House: Cancer Activists, Discourses of Prevention, and Environmental Justice." *Identities* 5:183–217.

Appadurai, Arjun. 1996. *Modernity at Large: Cultural Dimensions of Globalization.* Minneapolis: University of Minnesota Press.

Arp, W., and K. Boeckelman. 1997. "Religiosity: A Source of Black Environmentalism and Empowerment?" *Journal of Black Studies* 28:255–268.

Augusta Chronicle. 1970. "3 Dead in Augusta Riots; State Guardsmen Activated." May 12, A:1, 10.

Auletta, Ken. 1982. *The Underclass.* New York: Random House.

Austin, Regina, and Michael Schill. 1994. "Black, Brown, Red and Poisoned." In *Unequal Protection.* Robert Bullard, ed. Pp. 53–76. San Francisco: Sierra Club Books.

Baer, Hans, and Yvonne Jones, eds. 1992. *African Americans in the South: Issues of Race, Class and Gender.* Athens: University of Georgia Press.

Baker, Houston. 2001. *Turning South Again: Rethinking Modernism/Re-reading Booker T. Washington.* Durham, N.C.: Duke University Press.

Baker, Lee. 1998. *From Savage to Negro: Anthropology and the Construction of Race, 1896–1954.* Berkeley: University of California Press.

Balibar, Etienne. 1991. "Is There a 'Neo-Racism'?" In *Race, Nation, Class: Ambiguous Identities.* Etienne Balibar and Immanuel Wallerstein, eds. Pp. 15–28. New York: Verso.

Bartholomew, Amy, and Margit Mayer. 1992. "Nomads of the Present: Melucci's Contribution to 'New Social Movement Theory.'" *Theory, Culture and Society* 9:141–159.

Basso, Keith. 1996. *Wisdom Sits in Places: Language and Land Use among the Western Apache.* Albuquerque: University of New Mexico Press.

Bayor, Ronald. 1988. "Roads to Racial Segregation: Atlanta in the Twentieth Century." *Journal of Urban History* 15:3–21.

———. 1996. *Race and the Shaping of Twentieth-Century Atlanta.* Chapel Hill: University of North Carolina Press.

Bebbington, Anthony. 1993. "Modernization from Below: An Alternative Indigenous Development?" *Economic Geography* 69:274–292.

Beck, Ulrich. 1992 [1986]. *Risk Society: Towards a New Modernity.* Trans. from the German by Mark Ritter, and with an introduction by Scott Lash and Brian Wynne. London: Sage.

Been, Vicki. 1993. "What's Fairness Got to Do with It? Environmental Justice and the Siting of Locally Undesirable Land Uses." *Cornell Law Review* 78: 1001–1005.

Been, Vicki, and F. Gupta. 1997. "Coming to the Nuisance or Going to the Barrio? A Longitudinal Analysis of Environmental Justice Claims." *Ecology Law Quarterly* 24:1–56.

Benford, Robert D., and David A. Snow. 2000. "Framing Processes and Social Movements: An Overview and Assessment." *Annual Review of Sociology* 26:611–639.

Berendt, John. 1994. *Midnight in the Garden of Good and Evil.* New York: Random House.

Berlin, Ira, and Philip D. Morgan. 1993. "Labor and the Shaping of Slave Life in the Americas." In *Cultivation and Culture: Labor and the Shaping of Slave Life in the Americas.* Ira Berlin and Philip D. Morgan, eds. Pp. 1–45. Charlottesville: University of Virginia Press.

Besson, Jean. 1995. "Women's Use of Roscas in the Caribbean: Reassessing the Literature." In *Money Go-Rounds: The Importance of Rotating Savings and Credit Associations for Women.* Shirley Ardner and Sandra Burman, eds. Pp. 263–288. Oxford: Berg.

Biles, Roger. 1986. *Memphis in the Great Depression.* Knoxville: University of Tennessee Press.

———. 1990. "The Urban South in the Great Depression." *Journal of Southern History* 56:71–100.

———. 1994. *The South and the New Deal.* Lexington: University of Kentucky Press.

Blu, Karen. 1980. *The Lumbee Problem: The Making of an American Indian People.* New York: Cambridge University Press.

Boas, Franz. 1938. *The Mind of Primitive Man.* New York: Free Press.

———. 1945. *Race and Democratic Society.* New York: J. J. Augustin.

Bosso, Christopher J. 1991. "Adaptation and Change in the Environmental Movement." In *Interest Group Politics.* Allan J. Cigler and Burdett A. Loomis, eds. Pp. 51–176. Washington, D.C.: CQ Press.

Brosius, J. Peter. 1999. "Green Dots, Pink Hearts: Displacing Politics from the Malaysian Rain Forest." *American Anthropologist* 101:36–57.

———. 2001. "The Politics of Ethnographic Presence: Sites and Topologies in the Study of Transnational Movements." In *New Directions in Anthropology and Environment: Intersections.* Carole Crumley, ed. Pp. 150–176. Walnut Creek, Calif.: AltaMira Press.

Brown, James, with Bruce Tucker. 1996. *James Brown: The Godfather of Soul.* New York: Thunder's Mouth Press.

Brown, Phil. 1992. "Toxic Waste Contamination and Popular Epidemiology: Lay and Professional Ways of Knowing." *Journal of Health and Social Behavior* 33:267–281.

Bryant, Bunyan. 1995a. "Introduction." In *Environmental Justice: Issues, Policies and Solutions*. Bunyan Bryant, ed. Pp. 1–7. Washington, D.C.: Island Press.

———. 1995b. "Pollution Prevention and Participatory Research as a Methodology for Environmental Justice." *Virginia Environmental Law Review* 14: 589–611.

Bryner, Gary. 2001. *Gaia's Wager: Environmental Movements and the Challenge of Sustainability*. New York: Rowman and Littlefield.

Buechler, Steven M. 1995. "New Social Movement Theories." *Sociological Quarterly* 36:441–464.

Bullard, Robert. 1989. *In Search of the New South: The Black Urban Experience in the 1970s and 1980s*. Tuscaloosa: University of Alabama Press.

———. 1993. *Confronting Environmental Racism: Voices from the Grassroots*. Boston: South End Press.

———. 1994. *Unequal Protection: Environmental Justice and Communities of Color*. San Francisco: Sierra Club Books.

———. 2000. *Dumping in Dixie: Race, Class and Environmental Quality*. 3d edition. Boulder, Colo.: Westview Press.

Bullard, Robert, and Beverly Wright. 1987. "Blacks and the Environment." *Humboldt Journal of Social Relations* 14:165–184.

Bunche, Ralph J. 1973. *The Political Status of the Negro in the Age of FDR*. Chicago: University of Chicago Press.

Burdick, John. 1998. *Blessed Anastácia: Women, Race, and Popular Christianity in Brazil*. New York: Routledge.

Caldwell, Erskine. 1932. *Tobacco Road*. New York: Scribner's.

Cardoso, Ruth. 1992. "Popular Movements in the Context of the Consolidation of Democracy in Brazil." In *The Making of Social Movements in Latin America*. Arturo Escobar and Sonia Alvarez, eds. Pp. 291–302. Boulder, Colo.: Westview.

Carson, Clayborne. 1995 [1981]. *In Struggle: SNCC and the Black Awakening of the 1960s*. Cambridge: Harvard University Press.

Carson, Rachel. 1962. *Silent Spring*. Boston: Houghton Mifflin.

Cash, Wilbur Joseph. 1941. *The Mind of the South*. New York: Knopf.

Cashin, Edward. 1980. *The Story of Augusta*. Augusta, Ga.: Richmond County Board of Education.

———. 1985. *The Quest: A History of Public Education in Richmond County, Georgia*. Augusta, Ga.: Richmond County Board of Education.

———. 2001. *Paternalism in a Southern City: Race, Religion, and Gender in Augusta, Georgia*. Athens: University of Georgia Press.

Castells, Manuel. 1997. *The Power of Identity*. Malden, Mass.: Blackwell.

Center for Policy Alternatives. 1994. *Toxic Wastes and Race Revisited*. Washington, D.C.: Center for Policy Alternatives.

Chamber of Commerce, Augusta, Georgia. 1947. "A Brief on Industrial Augusta, Georgia."

Chavis, Benjamin. 1993. "Foreword." In *Confronting Environmental Racism: Voices from the Grassroots.* Robert Bullard, ed. Pp. 3–5. Boston: South End.

Checker, Melissa. 2001. "Like Nixon Coming to China: Finding Common Ground in a Multi-Ethnic Coalition for Environmental Justice." *Anthropological Quarterly* 74:135–146.

———. 2002a. "It's in the Air: Redefining the Environment as a New Metaphor for Old Social Justice Struggles." *Human Organization* 61:94–105.

———. 2002b. "Troubling the Waters: Race, Activism and the Environment in the U.S. South." Ph.D. dissertation, New York University.

———. 2004a. "Treading Murky Waters: Day-to-Day Dilemmas in the Construction of a Pluralistic U.S. Environmental Movement." In *Local Actions: Cultural Activism, Power, and Public Life in America.* Melissa Checker and Maggie Fishman, eds. Pp. 27–50. New York: Columbia University Press.

———. 2004b. "'We All Have Identity at the Table': Negotiating Difference in a Southern African American Environmental Justice Network." Identities: Global Studies in Culture and Power 11:171–194.

———. 2005. From Friend to Foe and Back Again: Industry and Environmental Action in the Urban South." *Urban Anthropology and Studies of Cultural Systems and World Economic Development* 34:7–44.

Checker, Melissa, and Maggie Fishman. 2004. "Introduction." In *Local Actions: Cultural Activism, Power, and Public Life in America.* Melissa Checker and Maggie Fishman, eds. Pp. 1–26. New York: Columbia University Press.

Cline, Damon. 1998. "Mission to Benefit Economy." *Augusta Chronicle,* December 23, 1A.

Cobb, James. 1975. "Politics in a New South City: Augusta, Georgia, 1946–1971." Ph.D. dissertation, University of Georgia.

———. 1982. "Yesterday's Liberalism: Boosterism and Racial Progress in Augusta, Georgia." In *Southern Businessmen and Desegregation.* David Coulburn and Elizabeth Jacoway, eds. Pp. 151–169. Baton Rouge: Louisiana State University Press.

———. 1999. *Redefining Southern Culture: Mind and Identity in the Modern South.* Athens: University of Georgia Press.

Cogan, John. 1982. "The Decline in Black Teenage Employment 1950–1970." *American Economic Review* 72:621–638.

Cohen, Jean. 1992. "Strategy or Identity: New Theoretical Paradigms and Contemporary Social Movements." *Social Research* 52:663–716.

Cole, Luke W., and Sheila R. Foster. 2001. *From the Ground Up: Environmental Racism and the Rise of the Environmental Justice Movement.* New York: NYU Press.

Coleman, Kenneth. 1978. *Georgia History in Outline.* Athens: University of Georgia Press.

Colopy, James. 1994. "The Road Less Traveled: Pursuing Environmental Justice through Title VI of the Civil Rights Act of 1964." *Stanford Environmental Law Journal* 13:125–189.

Commission for Racial Justice. 1987. Toxic Wastes and Race in the United States: A National Report on the Racial and Socioeconomic Characteristics of Communities with Hazardous Waste Sites. New York: United Church of Christ.

Conrad, D. E. 1965. *The Forgotten Farmers: The Story of Sharecroppers in the New Deal.* Champaign: University of Illinois Press.

Cooper, Sylvia. 1993. "Neighbors Near Wood Plant Get EPA Results." *Augusta Chronicle,* September 17, 15A.

———. 1998. "Race Flavors Mayor Picks." *Augusta Chronicle,* November 22, 1A, 6A.

Craig, Stewart. 1991a. "Contamination Spoils Garden." *Augusta Chronicle,* June 26, 1D, 4D.

———. 1991b. "Southern Wood Misled State on Toxins, Scientists Say." *Augusta Chronicle,* March 15, 1A, 12A.

Crumley, Carole, ed. 2001. *New Directions in Anthropology and the Environment: Intersections.* Walnut Creek, Calif.: AltaMira Press.

Cushing, Frank Hamilton. 1979. *Zuni: Selected Writings of Frank Hamilton Cushing.* Lincoln: University of Nebraska Press.

Cutter, Susan L., Michael S. Scott, and Arlene A. Hill. 2002. "Spatial Variability in Toxicity Indicators Used to Rank Chemicals." *American Journal of Public Health* 92:420–422.

Daniel, E. Valentine. 1984. *Fluid Signs: Being a Person the Tamil Way.* Berkeley: University of California Press.

Dark, Alx. 1998. "Public Sphere Politics and Community Conflict over the Environment and Native Land Rights in Clayoquot Sound, British Columbia." Ph.D. dissertation, New York University.

Darnovsky, Marcy. 1992. "Stories Less Told: Histories of U.S. Environmentalism." *Socialist Review* 22 (4):11–54.

Darnovsky, Marcy, Barbara Leslie Epstein, and Richard Flacks, eds. 1995. *Cultural Politics and Social Movements.* Philadelphia: Temple University Press.

Davies, A. F. 1980. *Skills, Outlooks and Passions.* Cambridge: Cambridge University Press.

Davis, Allison, Burleigh Gardner, and Mary Gardner. 1941. *Deep South.* Chicago: University of Chicago Press.

di Leonardo, Micaela. 1998. *Exotics at Home: Anthropologies, Others, American Modernity.* Chicago: University of Chicago Press.

Donahue, Kathleen. 1992a. "Plant Neighbors Air Concerns." *Augusta Chronicle,* January 12, 1B, 5B.

———. 1992b. "Residents Ready for a Battle." *Augusta Chronicle,* January 17, 3A.

———. 1992c. "Southern Wood Reports Ruled Legal." *Augusta Chronicle,* July 10, 1A.

Douglas, Mary, and Aaron Wildavsky. 1982. *Risk and Culture: An Essay on the Selection of Technological and Environmental Dangers.* Berkeley: University of California Press.

Downs, George W. 2000. "Constructing Effective Environmental Regimes." *Annual Review of Political Science* 3:25–42.

Drake, St. Clair. 1987. *Black Folk Here and There: An Essay in Anthropology and History.* Volume 2. Los Angeles: Center for Afro-American Studies, University of California.

Drake, St. Clair, and Horace Cayton. 1945. *Black Metropolis: A Study of Negro Life in a Northern City.* Chicago: University of Chicago Press.

Dryzek, John. 1999. "Global Ecological Democracy." In *Global Ethics and Environment.* Nicholas Low, ed. Pp. 264–282. New York: Routledge.

Du Bois, W. E. B. 1896 [1903]. *The Souls of Black Folk.* New York: Vintage.

———. 1935. *Black Reconstruction in America: An Essay toward a History of the Part Which Black Folk Played in the Attempt to Reconstruct Democracy, 1860–1880.* New York: Harcourt, Brace.

Dudziak, Mary L. 2000. *Cold War Civil Rights: Race and the Image of American Democracy.* Princeton: Princeton University Press.

Duneier, Mitchell. 1998. *Slim's Table: Race, Respectability, and Masculinity.* Chicago: University of Chicago Press.

Eddy, Elizabeth M., and William L. Partridge. 1978. "The Development of Applied Anthropology in America." In *Applied Anthropology in America.* Elizabeth M. Eddy and William L. Partridge, eds. Pp. 3–55. New York: Columbia University Press.

Edelman, Marc. 2001. "Social Movements: Changing Paradigms and Forms of Politics." *Annual Review of Anthropology* 30:285–317.

Edelstein, Michael. 2003. *Contaminated Communities: Coping with Residential Toxic Exposure.* Boulder, Colo.: Westview.

Enloe, Cynthia. 1989. *Bananas, Beaches and Bases.* Berkeley: University of California Press.

Escobar, Arturo. 1992. "Culture, Economics, and Politics in Latin American Social Movements Theory and Research." In *The Making of Social Movements in Latin America.* Arturo Escobar and Sonia Alvarez, eds. Pp. 62–88. Boulder, Colo.: Westview.

———. 1995. *Encountering Development: The Making and Un-making of the Third World.* Princeton: Princeton University Press.

Escobar, Arturo, and Sonia Alvarez. 1992. *The Making of Social Movements in Latin America.* Boulder, Colo.: Westview.

Faber, Daniel, ed. 1998. *The Struggle for Ecological Democracy: Environmental Justice Movements in the United States.* New York: Guilford.

Fisher, William F. 1997. "Doing Good? The Politics and Anti-politics of NGO Practices." *Annual Review of Anthropology* 26:439–464.

Fitchen, Janet. 1988. "Anthropology and Environmental Problems in the U.S.: The Case of Groundwater Contamination." *Practicing Anthropology* 10 (3–4): 5, 18.

Foley, Douglas E. 1999. "The Fox Project: A Reappraisal." *Current Anthropology* 40:171–191.

Fordham, Signithia. 1996. *Blacked Out: Dilemmas of Race, Identity and Success at Capital High.* Chicago: University of Chicago Press.

Forman, James. 1972. *The Making of Black Revolutionaries.* New York: Macmillan.

Franklin, Sara. 1995. "Science as Culture, Culture as Science." *Annual Review of Anthropology* 24:163–184.

Frazier, E. Franklin. 1939. *The Negro Family in the United States.* Chicago: University of Chicago Press.

———. 1963. *The Negro Church in America.* New York: Schocken Books.

Gaines, Kevin. 1996. *Uplifting the Race: Black Leadership, Politics, and Culture in the Twentieth Century.* Chapel Hill: University of North Carolina Press.

Gatewood, Willard B., Jr. 1990. *Aristocrats of Color: The Black Elite, 1880–1920.* Fayetteville: University of Arkansas Press.

Geisst, Charles. 2000. *Monopolies in America: Empire Builders and Their Enemies from Jay Gould to Bill Gates.* New York: Oxford University Press.

Genovese, Eugene D. 1976. *Roll, Jordan, Roll: The World That Slaves Made.* New York: Pantheon.

Gerlach, Luther, and Virginia Hine. 1970. *People, Power, Change: Movements of Social Transformation.* New York: Bobbs-Merrill.

Gezon, Lisa. 2000. "Anthropologists and NGOs—The Changing Face of NGOs: Structure and Communitas in Conservation and Development in Madagascar." *Urban Anthropology and Studies of Cultural Systems and World Economic Development* 29:181–215.

Gilroy, Paul. 2000. *Against Race: Imagining Political Culture Beyond the Color Line.* Cambridge: Harvard University Press.

Ginsburg, Faye, and Anna Lowenhaupt Tsing, eds. 1990. *Uncertain Terms: Negotiating Gender in American Culture.* Boston: Beacon.

Giugni, Marco G. 1998. "Was It Worth the Effort? The Outcomes and Consequences of Social Movements." *Annual Review of Sociology* 24:371–393.

Glazer, Nathan, and Daniel P. Moynihan. 1963. *Beyond the Melting Pot: The Negroes, Puerto Ricans, Jews, Italians, and Irish of New York City.* Cambridge: MIT Press.

Goldberg, Robert. 1991. *Grassroots Resistance: Social Movements in the Twentieth Century.* Prospect Heights, Ill.: Waveland.

Goldfield, David R. 1987. *Promised Land: The South since 1945.* Arlington Heights, Ill.: Harlan Davidson.

Goldman, Michael, and Rachel A. Schurman. 2000. "Closing the 'Great Divide': New Social Theory on Society and Nature. *Annual Review of Sociology* 26:563–584.

Goldschmidt, Henry. 2004. "Food Fights: Contesting 'Cultural Diversity' in Crown Heights." In *Local Actions: Cultural Activism, Power and Public Life in America.* Melissa Checker and Maggie Fishman, eds. Pp. 159–183. New York: Columbia University Press.

Goodwin, Jeff, and James Jasper. 1999. "Caught in a Winding, Snarling Vine: The Structural Bias of Political Process Theory." *Sociological Forum* 14:29–54.

———, eds. 2003. *Rethinking Social Movements: Structure, Meaning, and Emotions.* Landham, Md.: Rowman and Littlefield.

Goodwin, Jeff, James M. Jasper, and Francesca Polletta, eds. 2001. *Passionate Politics: Emotions and Social Movements.* Chicago: University of Chicago Press.

Gottlieb, Robert. 1993. *Forcing the Spring: The Transformation of the American Environmental Movement.* Washington, D.C.: Island Press.

Gottlieb, Robert, and Hal Ingram. 1988. "Which Way Environmentalism? Towards a New Movement." In *Winning American Education.* C. Hartman and M. Raskin, eds. Pp. 114–130. Boston: South End.

Governor's Task Force Report. 1996. *Governor's Task Force Report on the Long-Term Health Care Needs for Southern Wood Piedmont Residents.* Augusta, Ga. Author's files.

Gramsci, Antonio. 1991. *The Prison Notebooks.* Joseph Buttigieg, ed. Joseph Buttigieg and Antonio Callari, trans. New York: Columbia University Press.

Gregory, Steven. 1992. "The Changing Significance of Race and Class in an African-American Community." *American Ethnologist* 19:255–274.

———. 1994. "We've Been Down This Road Already." In *Race.* Steven Gregory and Roger Sanjek, eds. Pp. 18–40. New Brunswick: Rutgers University Press.

———. 1998. *Black Corona: Race and the Politics of Place in an Urban Community.* Princeton: Princeton University Press.

Griffith, David Craig, and Manuel Valdez-Pizzini. 2002. *Fishers at Work, Workers at Sea: A Puerto Rican Journey through Labor and Refuge.* Philadelphia: Temple University Press.

Grossman, James R. 1989. *Land of Hope: Chicago, Black Southerners, and the Great Migration.* Chicago: University of Chicago Press.

Grunwald, Michael. 2002. "Monsanto Hid Decades of Pollution." *Washington Post,* January 1, A1.

Gwaltney, John Langston. 1993. *Drylongso: A Self-Portrait of Black America.* New York: New Press.

Haddock, Brandon. 1999. "Disposal Mission to Bring Plutonium Full Circle." *Augusta Chronicle,* May 30, 3.

———. 2000. "Masters Week Brings Change." *Augusta Chronicle,* February 4, 1B.

Haenn, Nora. Forthcoming. "Risking Environmental Justice: Culture, Conservation, and Governance at Calakmul, Mexico." In *Social Injustice in Latin America,* Susan Eckstein and Timothy Wickham-Crawley, eds. New York: Routledge.

Halbwachs, Maurice. 1992. *On Collective Memory.* Chicago: University of Chicago Press.

Hall, Robert, and Carol Stack, eds. 1982. *Holding onto the Land and the Lord: Essays on Kinship, Ritual, Land Tenure and Social Policy.* Athens: University of Georgia Press.

Halperin, Rhoda H. 1998. *Practicing Community: Class Culture and Power in an Urban Neighborhood.* Austin: University of Texas Press.

Hamilton, Cynthia. 1990. "Women, Home and Community: The Struggle in an Urban Environment." *Race, Poverty and the Environment* 1 (1): 3, 10–13.

Hannerz, Ulf. 1969. *Soulside: Inquiries into Ghetto Culture and Community.* New York: Columbia University Press.

Hansell, Saul. 1993. "Shamed by Publicity, Banks Stress Minority Mortgages," *New York Times,* August 30, D1, D2.

Haraway, Donna. 1989. *Primate Visions: Gender, Race and Nature in the World of Modern Science.* New York: Routledge.

———. 1991. *Simians, Cyborgs, and Women: The Reinvention of Nature.* London: Free Association.

Harrison, Faye. 1995a. "'Give Me That Old-Time Religion': The Genealogy and Cultural Politics of an Afro-Christian Celebration in Halifax County, North Carolina." In *Religion in the Contemporary U.S. South: Diversity, Community, Identity.* D. Kendall White Jr. and Daryl White, eds. Pp. 34–45. Athens: University of Georgia Press.

———. 1995b. "The Persistent Power of 'Race' in the Cultural and Political Economy of Racism." *Annual Review of Anthropology* 24:47–74.

———. 1998. "Introduction: Expanding the Discourse on 'Race.'" *American Anthropologist* 100:609–631.

Harvey, David. 1989. *The Condition of Postmodernity: An Enquiry into the Origins of Cultural Change.* Cambridge, Mass.: Blackwell.

———. 1996. "The Environment of Justice." In *Justice, Nature and the Geography of Difference.* Cambridge, Mass.: Blackwell.

Henri, Florette. 1975. *Black Migration: Movement North, 1900–1920.* Garden City, N.Y.: Anchor/Doubleday.

Herskovitz, Melville. 1990 [1941]. *The Myth of the Negro Past.* Boston: Beacon.

Hewell, Hal. 1989. "Samples Collected at Park." *Augusta Herald,* June 24, 1A, 8A.

Higginbotham, Evelyn Brooks. 1993. *Righteous Discontent: The Women's Movement in the Black Baptist Church, 1880–1920.* Cambridge: Harvard University Press.

Hill, Carole E. 1998. "Contemporary Issues in Anthropological Studies of the American South." In *Cultural Diversity in the U.S. South: Anthropological Contributions to a Region in Transition.* Carole E. Hill and Patricia D. Beaver, eds. Pp. 12–33. Athens: University of Georgia Press.

Hirsch, Arnold, and Raymond Mohl, eds. 1993. *Urban Policy in Twentieth-Century America.* New Brunswick: Rutgers University Press.

Holton, Tom. 1969. "Councilman 'Detained'; Sheriff Orders Probe." *Augusta Chronicle,* December 18, A1.

Hornblower, Margo. 1997. "Generation X Gets Real—The Eco-Supremo: Adam Werbach, 24." *Time,* June 9, 1997, 66–67.

Hochschild, Jennifer. 1996. *Facing Up to the American Dream: Race, Class and the Soul of the Nation.* Princeton: Princeton University Press.

Huth, Elizabeth. 1991a. "Dow Wants Dismissal." *Augusta Herald,* July 16, 1A, 7A.

———. 1991b. "Guidelines Shed Light on Wood-Plant Soil Tests." *Augusta Herald,* July 20, 1A, 7A.

Hyland, Stan, Ronald Register, and Kristine Gunther. 1991. "The Role of Cities in the Economic Development of the Lower South." *City and Society* 5:2.

Ingold, Tim. 1997. "Eight Themes in the Anthropology of Technology." *Social Analysis* 41:106–138.

Institute for Southern Studies. 1992. "States with High African American and Hispanic Populations Rank Worst on Environmental Health, According to 1991–1992 Green Index Report." News release, April 18.

Israel, Brian. 1995. "An Environmental Justice Critique of Risk Assessment." *New York University Environmental Law Journal* 3:469–522.

Jackson, John. 2001. *Harlemworld: Doing Race and Class in Contemporary Black America.* Chicago: University of Chicago Press.

Jackson, Thomas. 1993. "The State, the Movement, and the Urban Poor: The War on Poverty and Political Mobilization in the 1960s." In *Urban Policy in Twentieth-Century America.* Arnold Hirsch and Raymond Mohl, eds. Pp. 403–449. New Brunswick: Rutgers University Press.

Jameson, Frederic. 1991. *Postmodernism, or the Cultural Logic of Late Capitalism.* Durham, N.C.: Duke University Press.

Jargowsky, Paul. 1997. *Poverty and Place: Ghettos, Barrios and the American City.* New York: Russell Sage.

Jasanoff, Sheila. 1993. "What Judges Should Know about the Sociology of Science." *Judicature* 77:77–82.

Jasanoff, Sheila, Gerald Markle, James Peterson, and Trevor Pinch, eds. 1995. *Handbook of Science and Technology Studies*, rev. edition. Thousand Oaks, Calif.: Sage.

Jasper, James M. 1997. *The Art of Moral Protest: Culture, Biography, and Creativity in Social Movements.* Chicago: University of Chicago Press.

———. 1998 "The Emotions of Protest: Affective and Reactive Emotions in and around Social Movements." *Sociological Forum* 13:397–424.

Johnson, Charles. 1941. *Growing Up in the Black Belt: Negro Youth in the Rural South.* Washington, D.C.: American Council on Education.

Johnson, Michelle Susan. 1994. "Juju Leaves in the Center of a Whirlwind: African American Nature/Culture Mediation." Ph.D. dissertation, University of Michigan.

Johnson Reagon, Bernice. 1993. *We'll Understand It Better By and By: Pioneering African American Gospel Composers.* Washington, D.C.: Smithsonian Books.

Johnston, Barbara Rose, ed. 1994. *Who Pays the Price?* Washington, D.C.: Island Press.

———. 1997. *Life and Death Matters: Human Rights and the Environment at the End of the Millennium.* Walnut Creek, Calif.: AltaMira Press.

———. 1998. *Water, Culture and Power.* John Donahue and Barbara Rose Johnston, eds. Washington, D.C.: Island Press.

Jones, Jacqueline. 1993. "Southern Diaspora: Origins of the Northern 'Underclass.'" In *The "Underclass" Debate: Views from History.* Michael B. Katz, ed. Pp. 27–54. Princeton: Princeton University Press.

Jordan, Winthrop. 1968. *White over Black: American Attitudes towards the Negro 1550–1812.* New York: Norton.

Katz, Michael B., ed. 1993. *The "Underclass" Debate: Views from History.* Princeton: Princeton University Press.

Kelley, Robin D. G. 1990. *Hammer and Hoe: Alabama Communists during the Great Depression.* Chapel Hill: University of North Carolina Press.

———. 1994. *Race Rebels: Culture, Politics and the Black Working Class.* New York: Free Press.

———. 1997. *Yo' Mama's Disfunktional! Fighting the Culture Wars in Urban America.* Boston: Beacon.

Kirby, Jack Temple. 1987. *Rural Worlds Lost: The American South, 1920–1960.* Baton Rouge: Louisiana State University Press.

Kirby, John B. 1980. *Black Americans in the Roosevelt Era: Liberalism and Race.* Knoxville: University of Tennessee Press.

Kirschenman, Joleen, and Kathryn M. Neckerman. 1991. "'We'd Love to Hire Them, But . . .': The Meaning of Race for Employers." In *The Urban Under-*

class. Christopher Jencks and Paul Peterson, eds. Pp. 203–232. Washington, D.C.: Brookings Institution.

Krasniewicz, Louise. 1992. *Nuclear Summer: The Clash of Communities at the Seneca Women's Peace Encampment*. Ithaca, N.Y.: Cornell University Press.

Kroll-Smith, Steven, and Hugh Floyd. 1997. *Bodies in Protest: Environmental Illness and the Struggle over Medical Knowledge*. New York: NYU Press.

Ladner, Joyce. 1971. *Tomorrow's Tomorrow: The Black Woman*. New York: Doubleday.

Langford, James. 1993. "EPA's Hyde Park Tests Expected to Be Conclusive." *Augusta Chronicle*, February 6, 13A.

Larana, Enrique, Hank Johnson, and Joseph Gusfield. 1994. *New Social Movements: From Ideology to Identity*. Philadelphia: Temple University Press.

Lassiter, Luke Eric. 2003. "Theorizing the Local." *Anthropology News* 44 (5): 13.

Latour, Bruno. 2004. "The Last Critique." *Harper's*, April, 15–20.

Lavelle, M., and M. Coyle. 1992. "Unequal Protection." *National Law Journal*, September 21, S1–12.

Lawrence, Charles R., III. 1987. "The Id, the Ego and Equal Protection: Reckoning with Unconscious Racism." *Stanford Law Review* 39:317–388.

Lazarus, Richard. 1993. "Pursuing 'Environmental Justice': The Distributional Effects of Environmental Protection." *Northwestern University Law Review* 87:787–856.

LeMaistre, Lisa. 1988. "In Search of a Garden: African-Americans and the Land in Piedmont Georgia." Master's thesis, University of Georgia, Athens.

Lemann, Nicholas. 1986. "The Origins of the Underclass, Pt. I." *Atlantic Monthly*, June, 35, 41, 47.

———. 1991. *The Promised Land: The Great Black Migration and How It Changed America*. New York: Knopf.

Lewan, Todd, and Dolores Barclay. 2001. "Torn from the Land." Electronic document. www.ap.org/pages/whatsnew/how.html.

Lewis, Earl. 1991. "Expectations, Economic Opportunities and Life in the Industrial Age: Black Migration to Norfolk, VA. 1910–1945." In *The Great Migration in Historical Perspective*. Joe William Trotter, ed. Pp. 22–45. Bloomington: Indiana University Press.

Lewis, Oscar. 1966. "The Culture of Poverty." *Scientific American* 215:19–25.

Lichterman, Paul. 2002. "Seeing Structure Happen: Theory-Driven Participant Observation." In *Methods of Social Movement Research*. Bert Klandermans and Suzanne Staggenborg, eds. Pp. 3–31. Minneapolis: University of Minnesota Press.

Liebow, Edward. 1988. "Permanent Storage for Nuclear Power Plant Wastes: Comparing Risk Judgments and Their Social Effects." *Practicing Anthropology* 10 (3–4): 10–12.

Liebow, Elliot. 1967. *Tally's Corner: A Study of Negro Streetcorner Men.* Boston: Little, Brown.

Light, Ivan. 1972. *Ethnic Enterprise in America.* Berkeley: University of California Press.

Lincoln, C. Eric, and Lawrence Mamiya. 1990. *The Black Church in the African American Experience.* Durham, N.C.: Duke University Press.

Lipsitz, George. 1995. *A Life in the Struggle: Ivory Perry and the Culture of Opposition.* Philadelphia: Temple University Press.

Little, Paul. 1999. "Environments and Environmentalisms in Anthropological Research: Facing A New Millennium." *Annual Review of Anthropology* 28: 253–284.

Lowe, Lisa. 1996. *Immigrant Acts: On Asian American Cultural Politics.* Durham, N.C.: Duke University Press.

Lutz, Catherine. 2002. *Homefront: A Military City and the American Twentieth Century.* Boston: Beacon.

Lynch, Owen. 1994. "Urban Anthropology, Postmodernism, and Perspectives." *City and Society* 7:35–52.

———. 1996. "Contesting and Contested Identities: Manthura's Chaubes." In *Narratives of Agency: Self-Making in China, India and Japan.* Wimal Dissanayake, ed. Pp. 74–103. Minneapolis: University of Minnesota Press.

Mahon, Maureen. 1997. "The Black Rock Coalition and the Cultural Politics of Race in the United States." Ph.D. dissertation, New York University.

Marable, Manning. 1983. *How Capitalism Underdeveloped Black America: Problems in Race, Political Economy, and Society.* Updated edition. Boston: South End.

Marcus, George E., and Michael M. J. Fisher. 1986. *Anthropology as Cultural Critique.* Berkeley: University of California Press.

Markowitz, Lisa. 2001. "Finding the Field: Notes on the Ethnography of NGOs." *Human Organization* 6:40–46.

Martin, Emily. 1994. *Flexible Bodies: Tracking Immunity in American Culture —From the Days of Polio to the Age of AIDS.* Boston: Beacon.

Marx, Gary. 1979. *Protest and Prejudice: A Study of Belief in the Black Community.* Westport, Conn.: Greenwood Press.

Massey, Douglas, and Nancy A. Denton. 1993. *American Apartheid: Segregation and the Making of the Underclass.* Cambridge: Harvard University Press.

Mayo, Marjorie, and Gary Craig. 1995. "Community Participation and Empowerment: The Human Face of Structural Adjustment or Tools for Democratic Transformation." In *Community Empowerment: A Reader in Participation and Development.* Gary Craig and Marjorie Mayo, eds. Pp. 1–11. Atlantic Highlands, N.J.: Zed Books.

McAdam, Doug. 1982. *Political Process and the Development of Black Insurgency.* Chicago: University of Chicago Press.

McAdam, Doug, John D. McCarthy, and Mayer N. Zald. 1996. "Introduction: Opportunities, Mobilizing Structures, and Framing Processes—Toward a Synthetic, Comparative Perspective on Social Movements." In *Comparative Perspectives on Social Movements: Political Opportunities, Mobilizing Structures, and Cultural Framings*. Doug McAdam, John D. McCarthy, and Mayer N. Zald, eds. Pp. 1–20. New York: Cambridge University Press.

McAdam, Doug, Sidney Tarrow, and Charles Tilly. 2001. *Dynamics of Contention*. Cambridge: Cambridge University Press.

McCarthy, John D., and Mayer Zald. 1977. "Resource Mobilization and Social Movements: A Partial Theory." *American Journal of Sociology* 82:1212–1241.

McCoy, Carl Levert. 1984. "A Historical Sketch of Black Augusta, Georgia." Master's thesis, University of Georgia.

McDonogh, Gary. 1993. *Black and Catholic in Savannah, Georgia*. Knoxville: University of Tennessee Press.

McLaren, P. 1994. "Multiculturalism and the Postmodern Critique: Toward a Pedagogy of Resistance and Transformation." In *Between Borders: Pedagogy and the Politics of Cultural Studies*. H. Giroux and P. McLaren, eds. Pp. 192–224. New York: Routledge.

McMillen, Neil R. 1990. *Dark Journey: Black Mississippians in the Age of Jim Crow*. Urbana: University of Illinois Press.

McWhorter, John. 2000. *Losing the Race: Self-Sabotage in Black America*. New York: Free Press.

Mead, Lawrence. 1986. *Beyond Entitlement: The Social Obligations of Citizenship*. New York: Free Press.

Melucci, Alberto. 1985. "The Symbolic Challenge of Contemporary Movements." *Social Research* 52:789–816.

———. 1988. "Getting Involved: Identity and Mobilization in Social Movements." In *International Social Movement Research: From Structure to Action—Comparing Social Movements Research across Cultures*. Hansperter Kriesi, Sidney Tarrow, and Bert Klandermans, eds. Pp. 329–348. London: JAI Press.

———. 1989. *Nomads of the Present: Social Movements and Individual Needs in Contemporary Society*. Philadelphia: Temple University Press.

———. 1994. "A Strange Kind of Newness: What's 'New' in New Social Movements?" In *New Social Movements: From Ideology to Identity*. Enrique Larana, Hank Johnson, and Joseph Gusfield, eds. Pp. 101–130. Philadelphia: Temple University Press.

———. 1998. "Third World or Planetary Conflicts?" In *Cultures of Politics/Politics of Cultures: Re-visioning Latin American Social Movements*. Sonia Alvarez, Evelina Dagnino, and Arturo Escobar, eds. Pp. 422–429. Boulder, Colo.: Westview.

Metro Courier. 1991. "Hyde, Aragon Park Residents Charge County with Neglect." March 13–19, 1A.
———. 1993. "Utley to Residents Near Railroad Tracks: 'Watch Out.'" April 21, 1, 3.
Miller, Daniel. 2000. *The Internet: An Ethnographic Approach*. New York: Berg.
Milton, Kay. 1995. *Environmentalism and Cultural Theory: Exploring the Role of Anthropology in Environmental Discourse*. New York: Routledge.
Moberg, Mark. 2001. "Co-opting Justice: Transformation of a Multiracial Environmental Coalition in Southern Alabama." *Human Organization* 60: 166–177.
———. 2002. "Erin Brockovich Doesn't Live Here: Environmental Politics and Responsible Care in Mobile County, Alabama." *Human Organization*. 61: 377–389.
Montague, Peter. 2003. "Environmental Justice Requires Precautionary Action." Testimony before the California Environmental Protection Agency Advisory Committee on Environmental Justice.
Montgomery, David. 1967. *Beyond Equality: Labor and the Radical Republicans 1862–1872*. New York: Knopf.
Moore, Henry. 1905. *Augusta: Her Commerce, Industries and Attractions*. Augusta, Ga.: Post E. Traveler's Protective Association.
Morgen, Sandra. 2002. *Into Our Own Hands: The Women's Health Movement in the United States, 1969–1990*. New Brunswick: Rutgers University Press.
Morris, Aldon. 1984. *The Origins of the Civil Rights Movement: Black Communities Organizing for Change*. New York: Free Press.
Morsey, Soheir. 1994. "Beyond the Honorary 'White' Classification of Egyptians: Societal Identity in Historical Context." In *Race*. Steve Gregory and Roger Sanjek, eds. Pp. 175–198. New Brunswick: Rutgers University Press.
Moss, Kary L. 1996. "Race and Poverty as a Tool in the Struggle for Environmental Justice." *Poverty and Race* 1:1–2.
Moynihan, Daniel Patrick. 1965. "The Negro Family: The Case for National Action." Report. Washington, D.C.: U.S. Department of Labor.
Mullings, Leith. 1979. "Ethnicity and Stratification in the Urban United States." *Annuals of the New York Academy of Science* 318:10–22.
Murray, Charles. 1984. *Losing Ground: American Social Policy*. New York: Basic Books.
Myrdal, Gunnar. 1944. *An American Dilemma: The Negro Problem and Modern Democracy*. New York: Harper and Brothers.
———. 1963. *Challenge to Affluence*. New York: Pantheon.
Nash, June. 1997. "The Fiesta of the Word: The Zapatista Uprising and Radical Democracy in Mexico." *American Anthropologist* 99:261–274.
National Urban League. 2004. *The State of Black America 2004*. New York: National Urban League.

Nelkin, Dorothy. 1987. *Selling Science.* New York: Freeman.

———, ed. 1992. *Controversies: Politics of Technical Decisions.* 3d edition. Newbury Park, Calif.: Sage.

Newman, Katherine. 1988. *Falling from Grace: Downward Mobility in the Age of Affluence.* Berkeley: University of California Press.

———. 1999. *No Shame in My Game: The Working Poor in the Inner City.* New York: Vintage.

Novotny, Patrick. 1995. "Where We Live, Work and Play: Reframing the Cultural Landscape of Environmentalism in the Environmental Justice Movement." *New Political Science* 23:61–78.

———. 1998. "Popular Epidemiology and the Struggle for Community Health in the Environmental Justice Movement." In *The Struggle for Ecological Democracy: Environmental Justice Movements in the United States.* Daniel Faber, ed. Pp. 137–158. New York: Guilford.

O'Brien, Mary. 2000. *Making Better Environmental Decisions: An Alternative to Risk Assessment.* Cambridge: MIT Press.

Office of Solid Waste and Emergency Response. 2002. "Brownfields." Electronic document. www.epa.gov/swerops/bf.

Ogbu, John. 1978. *Minority Education and Caste: The American System in Cross-Cultural Perspective.* New York: Academic Press.

Omi, Michael, and Howard Winant. 1994 [1986]. *Racial Formation in the United States.* New York: Routledge.

Ortiz, Paul Andrew. 2000. "'Like Water Covered the Sea': The African American Freedom Struggle in Florida, 1877–1920." Ph.D. dissertation, Duke University.

Paolisso, Michael, and R. Shawn Maloney. 2000. "Recognizing Farmer Environmentalism: Nutrient Runoff and Toxic Dinoflagellate Blooms in the Chesapeake Bay Region." *Human Organization* 59:209–221.

Pattillo-McCoy, Mary. 1999. *Black Picket Fences: Privilege and Peril among the Black Middle Class.* Chicago: University of Chicago Press.

Pavey, Robert. 1992. "Plant Neighbors Air Concerns." *Augusta Chronicle,* January 12, 1A.

———. 1993. "Residents Tell EPA Wood Plant Study Flawed, 'Keep Out.'" *August Chronicle,* February 19, 1A.

———. 1994. "Hyde Park Residents Blast Agencies." *Augusta Chronicle,* June 1, 9A, 11A.

———. 1998. "Cleanup Begins in Hyde Park." *Augusta Chronicle,* June 26, 1C, 7C.

———. 2001. "Closed Hearing Held to Help Set Church's Value." *Augusta Chronicle,* May 28, B1.

Pavey, Robert, and Faith Johnson. 1998. "Factories Spark Concern." *Augusta Chronicle,* September 21, 1A, 6A.

Payne, Charles. 1995. *I've Got the Light of Freedom: The Organizing Tradition and the Mississippi Freedom Struggle.* Berkeley: University of California Press.

Pellow, David N., and Lisa Sun-Hee Park. 2003. *The Silicon Valley of Dreams: Environmental Injustice, Immigrant Workers, and the High-Tech Global Economy.* New York: NYU Press.

Pena, Devon. 2002. "Endangered Landscapes and Disappearing Peoples? Identity, Place, and Community in Ecological Politics." In *The Environmental Justice Reader: Politics, Poetics, and Pedagogy.* Joni Adamson, Mei Mei Evans, and Rachel Stein, eds. Pp. 58–81. Tucson: University of Arizona Press.

Pena, Devon, and Maria Mondragon-Valdez. 1998. "The 'Brown' and the 'Green' Revisited: Chicanos and Environmental Politics in the Upper Rio Grand." In *The Struggle for Ecological Democracy: Environmental Justice Movements in The United States.* Daniel Faber, ed. Pp. 312–348. New York: Guilford.

Perin, Constance. 1977. *Everything in Its Place: Social Order and Land Use in America.* Princeton: Princeton University Press.

Pfaffenberger, Bryan. 1995. "'The Second Self' in a Third World Immigrant Community." *Ethnos* 60 (1–2): 59–80.

Philanthropy News Digest. 1999. "Tech Firms Fund Academies to Help Bridge Digital Divide." July 13. http://fdncenter.org/pnd/archives/19990713/002800.html.

Pichardo, Nelson A. 1997. "New Social Movements: A Critical Review." *Annual Review of Sociology* 23:411–430.

Piven, Frances Fox, and Richard Cloward. 1977. *Poor People's Movements: Why They Succeed, How They Fail.* New York: Pantheon.

Polletta, Francesca. 2002. *Freedom Is an Endless Meeting: Democracy in American Social Movements.* Chicago: University of Chicago Press.

Polletta, Francesca, and James M. Jasper. 2001. "Collective Identity and Social Movements." *Annual Review of Sociology* 27:283–305.

Powdermaker, Hortense. 1939. *After Freedom: A Cultural Study in the Deep South.* New York: Viking.

Prince, Sabiyha Robin. 2002. "Changing Places: Race, Class and Belonging in the 'New' Harlem." *Urban Anthropology* 31:5–35.

Pugh, Tony. 2002. "Report: Doctors Deny Black Patients Care They Need." *Orlando Sentinel,* A1, A12.

Pulido, Laura. 1996. *Two Chicano Struggles in the Southwest.* Tucson: University of Arizona Press.

———. 1998. "Ecological Legitimacy and Cultural Essentialism: Hispano Grazing in the Southwest." In *The Struggle for Ecological Democracy: Environmental Justice Movements in the United States.* Daniel Faber, ed. Pp. 293–311. New York: Guilford.

Quint, Michael. 1991. "Racial Gap Detailed Mortgages: Loan-Denial Rate Said to Be Double for Minorities." *New York Times,* October 22.

Rabin, Yale. 1990. "Expulsive Zoning: The Inequitable Legacy of Euclid." In *Zoning and the American Dream 101.* Charles Haar and Jerrold Kayden, eds. Chicago: Planner's Press.

Rahman, Muhammad Anisur. 1995. "Participatory Development: Toward Liberation or Co-optation?" In *Community Empowerment: A Reader in Participation and Development.* Gary Craig and Marjorie Mayo, eds. Pp. 24–32. Atlantic Highlands, N.J.: Zed Books.

Rangan, Haripriya. 1993. "Romancing the Environment: Popular Environmental Action in the Himalayas." In *The Defense of Livelihood: Comparative Studies on Social Action.* Ed Friedman and Haripriya Rangan, eds. Pp. 155–181. West Hartford, Conn.: Kumarian Press.

Raper, Arthur. 1936. *Preface to Peasantry: A Tale of Two Black Belt Counties.* Chapel Hill: University of North Carolina Press.

Reed, Adolph, Jr. 1986a. "The 'Black Revolution' and the Reconstitution of Domination." In *Race, Politics and Culture.* Adolph Reed Jr., ed. Pp. 61–95. Westport, Conn.: Greenwood.

———. 1986b. *Jesse Jackson and the Crisis of Purpose in Afro-American Politics.* New Haven, Conn.: Yale University Press.

———. 1988. "The Liberal Technocrat." *Nation* 246:167–170.

———. 1999. *Stirrings in the Jug: Black Politics in the Post-segregation Era.* Minneapolis: University of Minnesota Press.

Reed, Jean-Pierre. 2004. "Emotions in Context: Revolutionary Accelerators, Hope, Moral Outrage, and Other Emotions in the Making of Nicaragua's Revolution." *Theory and Society* 33 (6): 1–51.

Regan, R., and M. Legerton. 1990. "Economic Slavery or Hazardous Wastes? Robeson County's Economic Menu." In *Communities in Economic Crisis: Appalachia and the South.* J. Gaventa, B. E. Smith, and A. W. Willingham, eds. Pp. 146–157. Philadelphia: Temple University Press.

Reidy, Joseph P. 1993. "Obligation and Right: Patterns of Labor, Subsistence, and Exchange in the Cotton Belt of Georgia." In *Cultivation and Culture: Labor and the Shaping of Slave Life in the Americans.* Ira Berlin and Philip D. Morgan, eds. Pp. 138–154. Charlottesville: University Press of Virginia.

Ribeiro, Gustavo Lins. 1998. "Cybercultural Politics: Political Activism at a Distance in a Transnational World." In *Cultures of Politics/Politics of Cultures: Re-visioning Latin American Social Movements.* Sonia Alvarez, Evelina Dagnino, and Arturo Escobar, eds. Pp. 325–352. Boulder, Colo.: Westview.

Roberts, Bill. 2000. "Anthropologists and NGOs—NGO Leadership, Success, and Growth in Senegal: Lessons from Ground Level." *Urban Anthropology and Studies of Cultural Systems and World Economic Development* 29:143–180.

Roberts, J. Timmons, and Melissa M. Toffolon-Weiss. 2001. "Environmental Justice Struggles in Perspective." In *Chronicles from the Environmental Justice Frontline*. Pp. 1–28. New York: Cambridge University Press.

Roediger, David. 1991. *The Wages of Whiteness*. New York: Verso.

Rosaldo, Renato. 1993. *Culture and Truth: The Remaking of Social Analysis*. Boston: Beacon.

Rothman, Hal. 1988. *The Greening of a Nation? Environmentalism in the U.S. since 1945*. New York: Harcourt, Brace.

Rowland, A. Ray, and Helen Callahan. 1976. *Yesterday's Augusta*. Miami, Fla.: Seaman.

Rupp, Leila, and Verta Taylor. 1987. *Survival in the Doldrums: The American Women's Rights Movement, 1945 to the 1960s*. New York: Oxford University Press.

Rural Development Center. 1975. *Georgia County Guide*. Tifton, Ga.: Cooperative Extension Service.

———. 1985. *Georgia County Guide*. Tifton, Ga.: Cooperative Extension Service.

———. 1999. *Georgia County Guide*. Tifton, Ga.: Cooperative Extension Service.

Sacks, Karen Brodkin. 1988. *Caring by the Hour: Women, Work, and Organizing at Duke Medical Center*. Chicago: University of Chicago Press.

———. 1994. "How Did Jews Become White Folks?" In *Race*. Steven Gregory and Roger Sanjek, eds. Pp. 78–102. New Brunswick: Rutgers University Press.

Sanjek, Roger. 1994. "The Enduring Inequalities of Race." In *Race*. Steven Gregory and Roger Sanjek, eds. Pp. 1–18. New Brunswick: Rutgers University Press.

Satterfield, Terre. 1997. "*Voodoo Science* and Common Sense: Ways of Knowing Old-Growth Forests." *Journal of Anthropological Research* 53:443–459.

———. 2002. *Anatomy of a Conflict: Identity, Knowledge and Emotion in Old-Growth Forests*. East Lansing: Michigan State University Press.

Scheffer, Victor. 1991. *The Shaping of Environmentalism in America*. Seattle: University of Washington Press.

Schell, Lawrence M., and Melinda Denham. 2003. "Environmental Pollution in Urban Environments and Human Biology." *Annual Review of Anthropology* 32:111–134.

Schettler, Ted, Katherine Barrett, and Carolyn Raffensperger. 2002. "The Precautionary Principle: Protecting Public Health and the Environment." In *Life Support: The Environment and Human Health*. Michael McCally, ed. Pp. 239–256. Cambridge: MIT Press.

Schill, Karin. 1998. "Study: SRS Downsizing Hurt Economies." *Augusta Chronicle*, January 23, 1B.

Schrader-Frechette, Kirstin. 1999. "Chernobyl, Global Environmental Injustice and Mutagenic Threats." In *Life and Death Matters: Human Rights at the End of the Millennium.* Barbara Rose Johnston, ed. Pp. 70–89. Walnut Creek, Calif.: AltaMira Press.

Schulman, Bruce. 1991. *From Cottonbelt to Sunbelt: Federal Policy, Economic Development and the Transition of the South, 1938–1980.* New York: Oxford University Press.

Schwab, James. 1994. *Deeper Shades of Green: The Rise of Blue-Collar and Minority Environmentalism in America.* San Francisco: Sierra Club Books.

Scoones, I. 1999. "New Ecology and the Social Sciences: What Prospects for a Fruitful Engagement?" *Annual Review of Anthropology* 28:479–507.

Scott, Anne Firor. 1990. "Most Invisible of All: Black Women's Voluntary Associations. *Journal of Southern History* 56:3–22.

Shankar, Shalini. 2004 "Fobby or Tight? 'Multicultural Day' and Other Struggles in Two Silicon Valley High Schools." In *Local Actions: Cultural Activism, Power and Public Life in America.* Melissa Checker and Maggie Fishman, eds. Pp. 184–288. New York: Columbia University Press.

Shanklin, Eugenia. 1998. "The Profession of the Color Blind: Sociocultural Anthropology and Racism in the Twenty-First Century." *American Anthropologist* 100:669–680.

Silver, Christopher, and John V. Moeser. 1995. *The Separate City: Black Communities in the Urban South, 1940–1968.* Lexington: University Press of Kentucky.

Simpson, Andrea. 2002. "Who Hears Their Cry? African American Women and the Fight for Environmental Justice in Memphis, Tennessee." In *The Environmental Justice Reader: Politics, Poetics, and Pedagogy.* Joni Adamson, Mei Mei Evans, and Rachel Stein, eds. Pp. 82–104. Tucson: University of Arizona Press.

Singer, Merril. 2000. "Why I Am Not a Public Anthropologist." *Anthropology News,* September, 6–7.

Sitkoff, Harvard. 1978. *A New Deal for Blacks: The Emergence of Civil Rights as a National Issue, the Depression Decade.* New York: Oxford University Press.

Smoller, Jeff. 1970. "Someone Has to Talk to the Kids." *Augusta Chronicle,* May 13, A1.

Sociology Research Methods Students, Kim Davies, Robert Johnston, Ernestine Thompson, Robin Bengtson, Sarah Firman, Albert Jimenez, Karen Jones, and Regina Murray. 1998. *Hyde Park Neighborhood Survey Report.* Augusta: Department of Sociology, Augusta State University.

Stack, Carol. 1974. *All Our Kin: Strategies for Survival in a Black Community.* New York: Harper and Row.

———. 1996. *Call to Home: African Americans Reclaim the Rural South.* New York: Basic Books.

Stoffle, Richard W., Camilla L. Harshbarger, Florence V. Jensen, Michael J. Evans, and Paula Drury. 1988. "Risk Perception Shadows: The Superconducting Super Collider in Michigan." *Practicing Anthropology* 10 (3–4): 6–7.

Stonich, Susan, and Connor Bailey. 2000. "Resisting the Blue Revolution: Contending Coalitions Surrounding Industrial Shrimp Farming." *Human Organization* 59:23–36.

Strang, David, and Sarah A. Soule. 1998. "Diffusion in Organizations and Social Movements: From Hybrid Corn to Poison Pills." *Annual Review of Sociology* 24:265–290.

Strathern, Marilyn. 1997. "A Return to the Native." *Social Analysis* 41:15–28.

Sugrue, Thomas. 1993. "The Structures of Urban Poverty: The Reorganization of Space and Work in Three Periods of American History." In *The "Underclass" Debate: Views from History.* Michael B. Katz, ed. Pp. 85–117. Princeton: Princeton University Press.

Sullivan, Mercer. 1989. *"Getting Paid": Youth Crime and Work in the Inner City.* Ithaca, N.Y.: Cornell University Press.

Szasz, Andrew. 1994. *Ecopopulism: Toxic Waste and the Movement for Environmental Justice.* Minneapolis: University of Minnesota Press.

Takagi, Dana. 1994. "Post–Civil Rights Politics and Asian-American Identity: Admissions and Higher Education." In *Race.* Steven Gregory and Roger Sanjek, eds. Pp. 229–242. New Brunswick: Rutgers University Press.

Takaki, Ronald, ed. 2002. *Debating Diversity: Clashing Perspectives on Race and Ethnicity in America.* New York: Oxford University Press.

Tarlock, Dan. 1994. "City versus Countryside: Environmental Equity in Context." *Fordham Urban Law Journal* 21:461–491.

Tarrow, Sidney. 1998. *Power in Movement: Social Movements and Contentious Politics.* New York: Cambridge University Press.

Taylor, Dorceta. 1989. "Blacks and the Environment: Towards an Explanation of the Concern and Action Gap between Blacks and Whites." *Environment and Behavior* 21:175–205.

———. 1992. "Can the Environmental Movement Attract and Maintain the Support of Minorities?" In *Race and the Incidence of Environmental Hazards.* Bryant Bunyan and Paul Mohai, eds. Pp. 28–54. Boulder, Colo.: Westview.

Taylor, Verta. 1989. "Social Movement Continuity: The Women's Movement in Abeyance." *American Sociological Review* 54:761–775.

Terkel, Studs. 1992. *Race: How Blacks and Whites Think and Feel about the American Obsession.* New York: Doubleday.

Tesh, Sylvia Noble. 2000. *Uncertain Hazards: Environmental Activists and Scientific Proof.* Ithaca, N.Y.: Cornell University Press.

Time. 1977. "The American Underclass," August 29, 14–15.

Touraine, Alain. 1988 [1984]. *Return of the Actor: Social Theory in Postindustrial Society.* M. Godzich, trans. Minneapolis: University of Minnesota Press.

Townsend, Patricia. 2000. *Environmental Anthropology: From Pigs to Policies.* Long Grove, Ill.: Waveland Press.

Trotter, Joe William, Jr. 1985. *Black Milwaukee: The Making of an Industrial Proletariat, 1915–1945.* Champaign: University of Illinois Press.

———. 1991. *The Great Migration in Historical Perspective: New Dimensions of Race, Class and Gender.* Bloomington: Indiana University Press.

———. 1993. "Blacks in the Urban North: The 'Underclass Questions' in Historical Perspective." In *The "Underclass" Debate: Views from History.* Michael B. Katz, ed. Pp. 55–84. Princeton: Princeton University Press.

Turner, Terrence. 1993. "Anthropology and Multiculturalism: What Is Anthropology That Multiculturalists Should Be Mindful of It?" *Cultural Anthropology* 8:411–429.

United Church of Christ Commission for Racial Justice. 1987. *Toxic Wastes and Race in the United States: A National Report on the Racial and Socioeconomic Characteristics of Communities with Hazardous Waste Sites.* New York: United Church of Christ.

U.S. Bureau of the Census. 1940. *Population Report.* Washington, D.C.: U.S. Government Printing Office.

———. 2000. *Population Report.* Washington, D.C.: U.S. Government Printing Office.

U.S. Environmental Protection Agency, Risk Assessment Forum. 1986. "Guidelines for the Health Risk Assessment of Chemical Mixtures." *Federal Register* 51 (185): 34014–34025.

U.S. General Accounting Office. 1983. *Siting of Hazardous Waste Landfills and Their Correlation with Racial and Economic Status of Surrounding Communities.* Washington, D.C.: General Accounting Office.

Valentine, Charles A. 1968. *Culture and Poverty: Critique and Counterproposals.* Chicago: University of Chicago Press.

———. 1969. "Culture and Poverty: Critique and Counter-Proposals." *Current Anthropology* 19:181–201.

———. 1971. "'The Culture of Poverty': Its Scientific Implications and Its Implications for Action." In *The Culture of Poverty: A Critique.* Eleanor Leacock, ed. Pp. 193–225. New York: Simon and Schuster.

Wan, Lillian. 2003. "Black History in Augusta Timeline." Unpublished manuscript on file with author.

Warner, W. Lloyd. 1962. *American Life: Dream and Reality.* Chicago: University of Chicago Press.

Warschauer, Mark. 2003. *Technology and Social Inclusion: Rethinking the Digital Divide.* Cambridge: MIT Press.

Weiss, Nancy J. 1983. *Farewell to the Party of Lincoln: Black Politics in the Age of FDR.* Princeton: Princeton University Press.

West, Cornel. 1993. *Race Matters.* Boston: Beacon.

West, Patrick, Mark Fly, Frances Larkin, and Robert Marans. 1992. "Minority Anglers and Toxic Fish Consumptions: Evidence from a Statewide Survey in Michigan." In *Race and the Incidence of Environmental Hazards: A Time for Discourse*. Bunyan Bryant and Paul Mohai, eds. Pp. 100–113. Boulder, Colo.: Westview.

Westmacott, Richard. 1992. *African-American Gardens and Yards in the Rural South*. Knoxville: University of Tennessee Press.

Westra, Laura, and Peter Wenz, eds. 1995. *Faces of Environmental Racism: Confronting Issues of Global Justice*. Lanham, Md.: Rowman and Littlefield.

White, Renee T. 1999. *Putting Risk in Perspective: Black Teenage Lives in the Era of AIDS*. Lanham, Md.: Rowman and Littlefield.

Whiteley, P., and V. Masayesva. 1998. "The Use and Abuse of Aquifers: Can the Hopi Indians Survive Multinational Mining?" In *Water, Culture, and Power: Local Struggles in a Global Context*. J. M. Donahue and B. R. Johnston, eds. Pp. 9–34. Washington, D.C.: Island Press.

Wigley, Daniel, and Kristin Schrader-Frechette. 1996. "Environmental Racism and Biased Methods of Risk Assessment." *Risk: Health, Safety and Environment* 7:55–88.

Williams, Brett. 1992. "Commentary: Poverty among African Americans in the Urban United States." *Human Organization* 51:164–174.

———. 2001. "A River Runs through Us." *American Anthropologist* 103:409–431.

Williams, Dave. 2000. "Group OK's Rail Request." *Augusta Chronicle*, August 24, B2.

Williams, Melvin D. 1974. *Community in a Black Pentecostal Church: An Anthropological Study*. Pittsburgh: University of Pittsburgh Press.

Williams, Raymond. 1980. *Problems in Materialism and Culture: Selected Essays*. New York: Verso.

Wilson, William Julius. 1978. *The Declining Significance of Race: Blacks and Changing American Institutions*. Chicago: University of Chicago Press.

———. 1987. *The Truly Disadvantaged*. Chicago: University of Chicago Press.

———. 1996. *When Work Disappears: The World of the New Urban Poor*. New York: Knopf.

Wisner, Ben. 1997. "Environmental Justice, Health and Safety in Urban South Africa." In *Life and Death Matters: Human Rights at the End of the Millennium*. Barbara Rose Johnston, ed. Pp. 265–286. Walnut Creek, Calif.: AltaMira Press.

Wolfe, Amy. 1988. "Risk Communication: Who's Educating Whom?" *Practicing Anthropology* 10 (3–4): 13–14.

Wolters, Raymond. 1970. *Negroes and the Great Depression: The Problem of Economic Recovery*. Westport, Conn.: Greenwood.

Wright, Gavin. 1986. *Old South, New South: Revolutions in the Southern Economy since the Civil War.* New York: Norton.

———. 2001. "Afterword." In *The Second Wave: Southern Industrialization from the 1940s to the 1970s.* Philip Scranton, ed. Pp. 286–300. Athens: University of Georgia Press.

Young, John A. 2001. "Assessing Cooperation and Change: The SfAA and the EPA." *Practicing Anthropology* 23 (3): 47–49.

Zeff, Robbin L., Marsha Love, and Karen Stults. 1989. *Empowering Ourselves: Women and Toxics Organizing.* Arlington, Va.: Citizen's Clearinghouse for Hazardous Wastes.

Zimmerman, Rae. 1994. "Issues of Classification in Environmental Equity: How We Manage Is How We Measure." *Fordham Urban Law Journal* 21:633–669.

Index